D1588925

PERSONAL AND WIRELESS COMMUNICATIONS
Digital Technology and Standards

THE KLUWER INTERNATIONAL SERIES IN ENGINEERING AND COMPUTER SCIENCE

COMMUNICATIONS AND INFORMATION THEORY
Consulting Editor
Robert Gallager

Other books in the series:

PERSONAL AND WIRELESS COMMUNICATIONS
Digital Technology and Standards

by

Kun Il Park, Ph. D.
Bell Communications Research, Inc.

KLUWER ACADEMIC PUBLISHERS
Boston / Dordrecht / London

Distributors for North America:
Kluwer Academic Publishers
101 Philip Drive
Assinippi Park
Norwell, Massachusetts 02061 USA

Distributors for all other countries:
Kluwer Academic Publishers Group
Distribution Centre
Post Office Box 322
3300 AH Dordrecht, THE NETHERLANDS

Library of Congress Cataloging-in-Publication Data

A C.I.P. Catalogue record for this book is available
from the Library of Congress.

Printed on acid-free paper.

Printed in the United States of America

For Meyeon and Kyunja

CONTENTS

PREFACE   ix

1 INTRODUCTION   1

  1.1 General Characteristics of PCS   2

  1.2 PCS Spectrum Allocations   4

  1.3 Wireless Access Standards   5

  1.4 Organization of the Book   4

2 DIGITAL WIRELESS COMMUNICATIONS TECHNOLOGY   9

  2.1 What is a Digital Communications System?   9

  2.2 Advantages of Digital Communications Systems   12

  2.3 Main Technology Elements of Digital Wireless Communications Systems 13

  2.4 Source Coding   15

  2.5 Channel Coding   28

  2.6 Modulation   33

  2.7 Methods of Creating Physical Channels   38

3 THE STANDARD BASED ON THE NORTH AMERICAN HIGH-TIER TDMA SYSTEM   55

  3.1 Overview   55

  3.2 The Digital Control Channel   57

3.3 The Digital Traffic Channel   103

**4   THE STANDARD BASED ON THE NORTH AMERICAN
    HIGH-TIER CDMA SYSTEM**   123

4.1 Overview   123

4.2 Physical Channels   124

4.3 Logical Channels   131

4.4 Frame Formats   132

4.5 Digital Information Processing   146

4.6 Call Processing   154

4.7 Signaling Message Formats   159

**5   HIGHLIGHTS OF OTHER WIRELESS ACCESS
    STANDARDS**   177

5.1 The GSM-Based TDMA system   178

5.2 The Personal Access Communications System   197

5.3 The Wideband CDMA System   214

5.4 The Composite TDMA/CDMA System   215

5.5 The DECT-Based System   215

5.6 Wireless Access Standards for the Unlicensed Operation   216

REFERENCES   219

ABOUT THE AUTHOR   221

INDEX   223

PREFACE

Personal Communications Services (PCS) is a new concept which will expand the horizon of wireless communications beyond the limitations of current cellular systems to provide end users with the ability to communicate *"with anyone, anywhere, anytime."* In 1993, the United States Federal Communications Commission (FCC) allocated a total of 140 MHz of frequency bands in the vicinity of 2 GHz to be used for PCS. In 1994, the FCC started auctioning licenses to operate PCS in these frequency bands, and, at the time of this writing, the auction is continuing. With these two pivotal actions by the FCC, PCS has taken a big step toward its realization.

Because of the inherent nature of mobility, which is characteristic of personal communications, wireless communications and PCS have become inseparable concepts. In particular, PCS will critically depend on wireless technologies for the mobile-to-network access portion of the service, which is referred to as the common air interface. The topic of this book is the wireless access technology used for the common air interface in order to support PCS.

The purpose of this book is two-fold:

1. First, this book presents clear tutorial expositions of main digital technology elements employed for wireless access systems. The main objective here is to pull together all the important basic technical elements necessary in understanding wireless access systems at one place so that the reader can obtain an overall view of the technology quickly and systematically.

2. The second purpose of this book is to review the common air interface standards for PCS. There are literally thousands of pages devoted to the PCS common air interface standards. These standards documents are intended for experts who are familiar with not only the underlying technologies but also the process and

rationale behind the written words. These *reading-between-the-lines* aspects are not obvious in the written words of the standards and can be very cryptic to an average reader. With respect to reviewing the standards, this book has two main objectives. The first objective is to sift out and summarize important elements of the standards, which are buried in the veritable mountain of paper in all too unfamiliar terms, notations, and abbreviations. The second objective is to expand, almost paradoxically, some of the more important elements to explain the intent and significance of the written words of the standards. In order to accomplish the second objective, various means are used. Many supporting diagrams have been provided to illustrate important points and explain underlying concepts. In other instances, some of the diagrams and tables of the standards are expanded to provide details. The detailed discussion of the standards is provided using two main standards as examples, one TDMA standard, and one CDMA standard. For other standards, highlights of key elements of the standards are presented.

This book is written for two primary audiences:

1. *Managers and professionals involved in personal and wireless communications.* This book will provide them with a quick means of obtaining a comprehensive picture of overall aspects which are important in the area.

2. *Students and professors in the communications field.* This book can be used as a text for a one-semester wireless communications course in a graduate school curriculum of electrical engineering.

Kun Il Park
Red Bank, New Jersey

Dr. Park is an employee at Bell Communications Research, Inc. (Bellcore). The views expressed herein are those of Dr. Park, and do not necessarily represent the views of Bellcore or its owners.

1
INTRODUCTION

Wireless communications has a long history of commercial and military applications. In the recent past, however, wireless communications has witnessed an explosion of renewed interest from consumers for its application for mobile and personal communications. Cellular phones are now widely used by all types of consumers — business men and women, housewives, college students, and teenagers. As the demand for more sophisticated mobile and personal communications services increases, the existing cellular systems face limitations.

Personal Communications Services (PCS) is a new concept which will expand the horizon of wireless communications beyond the limitations of current cellular systems to provide users with the ability to communicate "*with anyone, anywhere, anytime.*" In 1993, the United States Federal Communications Commission (FCC) allocated frequency bands for PCS. In 1994, the FCC started auctioning licenses to operate PCS in these frequency bands, and, at the time of this writing, the auction is continuing. With these two pivotal actions by the FCC, PCS has taken a big step toward its realization.

Because of the inherent nature of mobility associated with personal communications, wireless communications and PCS have now become almost inseparable concepts. PCS will critically depend on, among other things, wireless technologies, particularly for the mobile-to-network access portion of the service, which is referred to as the common air interface. The topic of this book is the wireless access technology used for the common air interface in order to support PCS.

This book first presents clear tutorial expositions of essential building blocks of the digital wireless communications technology. It then provides in-depth examples of two major wireless access systems which have recently been standardized to support PCS: a Time Division Multiple Access (TDMA) system based on the North American TDMA cellular standard and a Code Division Multiple Access (CDMA) system based on the North American CDMA cellular standard. After these in-depth illustrations of the two standard wireless access systems, this book presents highlights of other wireless access standards for PCS.

In this introductory chapter, we first discuss the general characteristics of PCS and the FCC's spectrum allocations and auctioning of PCS licenses.

1.1 GENERAL CHARACTERISTICS OF PCS

The following general characteristics are often associated with PCS in order to provide *communications with anyone, anywhere, anytime*:

- *Wide user roaming ability.* "Roaming" refers to a user's ability to originate and receive calls while he/she is in other carriers' territory. In order to provide the roaming service, the user's carrier and the other carrier must have a business agreement to provide such a service. For PCS, the roaming service should be greatly expanded to provide for universal accessibility.

- *Wireless/wireline integration.* A wireless user may instruct the network to deliver the incoming call to a pre-designated wireline number. For example, the user may have a wireless phone at work and, as he/she leaves the office, may instruct the network (by a process known as "registration") to deliver the incoming call to his/her home phone number.

- *Diverse environments.* PCS users should be able to use the service in all types of environment: urban and rural business and residential areas, mountains and camping areas, recreational areas (e.g., golf courses, amusement parks), and large indoor areas such as indoor stadiums, airport buildings, and indoor shopping malls. The great variation in the type of environment implies wide variations in the radio propagation properties, which present more challenges to radio systems design engineers.

- *Various user mobility patterns.* The current cellular system was originally designed primarily for use in a vehicle. PCS will be used as much by pedestrians as inside moving vehicles.

■ *Portable handset with a long battery life.* A user should be able to carry the handset outside the car for a long time without having to recharge its battery. With the existing cellular phone system, the operating power level is relatively high and so the battery of the handset, albeit portable, cannot last too long before it needs to be recharged. For PCS, therefore, the battery life should be as long as possible. One way of achieving this is to use a low power. Of course, an alternative is to carry a large battery, which is not very desirable.

■ *Various power levels.* The existing cellular system operates basically at a high power level. Radio systems for PCS will have a variety of operating power levels.

■ *Various cell sizes.* Although individual cell sizes may vary somewhat depending on the type of system and applications, the existing cellular systems all have a generally large cell size. These systems are called macro cell systems. With PCS, there will be a mix of four broad types of cell sizes: the picocell for low power indoor applications using, for example, the unlicensed spectrum; the microcell for low power outdoor (or indoor) pedestrian applications in high population density areas; the macrocell for high power vehicular applications; and the super-macro cell with satellites. Use of satellites will be an essential element if the vision of communications "anywhere" is to be realized.

■ *Ubiquitous deployment of radio systems.* Use of low power micro cell systems requires a wide-spread deployment of radio systems. For microcell systems, therefore, unobtrusive small antennas are required for better community acceptance.

■ *The FCC's allocation of frequency spectrum.* One of the most significant elements which define PCS is the FCC's allocation of frequency bands to be used for PCS. As will be discussed in the next section, the FCC allocated a total of 120 MHz of spectrum for licensed operation and another 20 MHz for unlicensed operation, amounting to a total of 140 MHz for PCS. For comparison, the total bandwidth allocated for the existing cellular operation is 50 MHz (25 MHz each for the A- and B-band). In other words, the FCC has allocated for PCS about three times the spectrum currently used by the cellular network indicating its resolve to make PCS widely available.

1.2 PCS SPECTRUM ALLOCATIONS

The FCC's allocation of frequency bands for PCS are shown in Tables 1-1 and 1-2.

Block	Spectrum Low Side (MHz)	Spectrum High Side (MHz)	Bandwidth (MHz)
A	1850-1865	1930-1945	30
D	1865-1870	1945-1950	10
B	1870-1885	1950-1965	30
E	1885-1890	1965-1970	10
F	1890-1895	1970-1975	10
C	1895-1910	1975-1990	30
Total			120

Table 1.1   The PCS frequency bands for the licensed operation

Block	Spectrum (MHz)	Bandwidth (MHz)
Isochronous	1910-1920	10
Asynchronous	1920-1930	10
Total		20

Table 1.2   The PCS frequency bands for the unlicensed operation

For the licensed operation, three 30-MHz bands, namely, A, B and C bands, and three 10-MHz bands, namely, D, E, and F bands, have been allocated. For the unlicensed operation, a total of 20 MHz has been allocated. In this frequency band, no license is required to operate PCS. Out of the 20 MHz, 10 MHz is allocated for the isochronous operation (e.g., voice) and 10 MHz, for the asynchronous operation (e.g., data). Figure 1-1 shows the relative positions of the PCS frequency bands discussed above.

A	D	B	E	F	C	Unlicensed	A	D	B	E	F	C
15	5	15	5	5	15	20	15	5	15	5	5	15

Figure 1-1 The PCS frequency bands

The FCC's auction of PCS licenses is divided by two types of licensing areas: the Major Trading Areas (MTA's) and the Basic Trading Areas (BTA's). The United States is divided into 51 MTA's and 493 BTA's. Two 30-MHz blocks, A and B, are auctioned for the MTA licenses, and one 30-MHz block, C, and three 10-MHz blocks, D, E, and F, for the BTA licenses. Hence, in any geographic area of the U. S., there can be up to six competing operations using A, B, C, D, E and F block licenses in addition to the existing cellular operations.

The FCC's auction rules state that the maximum bandwidth that can be licensed by one PCS operator is 40 MHz including the current cellular bandwidth. If a prospective licensee currently operates a cellular service in an MTA, that operator cannot get a 30-MHz band (i.e., A, B or C band) because the current cellular system bandwidth is 25 MHz and an addition of 30 MHz block would make the total bandwidth for the operator 55 MHz violating the auction rule. The FCC completed the MTA auction, i.e., for A and B bands, on March 13, 1995. At the time of this writing, the next auction will be the 30-MHz block for BTA's, i.e., for the C band.

1.3 WIRELESS ACCESS STANDARDS

One essential step toward the realization of PCS is the standardization of the common air interface for the wireless access. Committee T1 is a standard setting organization which has primarily been interested in network standards. The participants of Committee T1 are typically representatives of local exchange carriers and inter-exchange carriers of the regulated industry and manufacturers supplying equipment for carriers. The Telecommunications Industry Associate (TIA) on the other hand represents the unregulated cellular industry and consists of typically the representatives of cellular carriers and cellular equipment manufacturers. Both Committee T1 and the TIA became interested in standardization of wireless access for PCS. These two standard setting bodies agreed to create a common body and, in 1992, formed the Joint Technical Committee (JTC).

The typical participants of the JTC include service providers (e.g., local exchange and inter-exchange carriers and cellular carriers), equipment manufacturers (e.g., radio, switch, test gear, antennas) and government representatives and individual consultants.   The scope of the JTC is to standardize the common air interface between the mobile station of the user and the base station of the wireless access system. The common air interface standards involve the Physical Layer, the Link Layer and the Network Layer.

Through most of 1993, the JTC developed a set of guidelines for the air interface standards. In November 1993, the JTC announced to the industry that the interested parties may submit their proposals to the JTC according to the guidelines developed by the JTC. The industry responded to this announcement quickly and a total of 16 proposals were submitted to the JTC by various companies. An obvious problem faced by the JTC was how to reduce this large number of proposals of vastly different systems into a single standard. At the very outset of the standardization process, the JTC realized that producing a single standard was not feasible, and, therefore, focused on reducing the 16 proposals as much as possible to a smaller number of standards.

The first task of the JTC was to group the 16 proposals into similar technologies. As one criterion for this grouping, the JTC distinguished between high-tier and low-tier systems. Broadly speaking, high-tier systems are designed for high power, large cell vehicular applications such as the existing cellular systems and low-tier systems are low power systems for small cell pedestrian applications. The other criterion used by the JTC for the grouping of the proposals was the multiple access method: the Time Division Multiple Access (TDMA), the Code Division Multiple Access (CDMA) and the composite TDMA/CDMA. Using these criteria, the JTC classified the 16 proposals into four major categories as follows:

- High-tier TDMA systems

    - Four proposals based on the existing European TDMA cellular standard, the Global System for Mobile (GSM)

    - Two proposals based on the existing North American TDMA cellular standard known as the IS-136 standard (formerly known as the IS-54 standard)

■ High-tier CDMA systems

  – Two proposals based on the existing North American CDMA cellular standard known as the IS-95 standard

  – Two wideband CDMA proposals, which are new and are not based on any existing standards

■ Low-tier TDMA systems

  – Two proposals based on the new Wireless Access Communications System (WACS), which are new and are not based on any existing standards

  – Two proposals based on the existing Japanese Personal Handy Phone (PHP) system

  – One proposal based on the existing Digital European Cordless Telephone (DECT) system

■ Composite CDMA/TDMA

  – One CDMA/TDMA composite system proposal, which is new and is not based on any existing standards

At the early stage of the JTC standardization process, an attempt was made to consolidate the proposals into a single standard per each of the four major categories. However, this attempt was not successful. The JTC then created eight Technical Adhoc Groups (TAG's) for the eight subgroups of the proposals as shown above by the dashed list. The participants of the eight TAG's agreed to combine the proposals into one within their respective TAG's and, by late December 1993, it became clear that the JTC could reduce the 16 proposals to eight standards. In January 1994, the WACS-based TAG and the PHP-based TAG agreed to merge their two proposals into one standard and named this combined standard the Personal Access Communications System (PACS). This WACS/PHP merger reduced the total number of standards to seven.

The final seven wireless access standards for PCS are:

■ One high-tier TDMA system based on the existing North American TDMA cellular standard, the IS-136 standard (formerly known as the IS-54)

- One high-tier CDMA system based on the existing North American CDMA cellular standard, the IS-95 standard

- One high-tier TDMA system based on the existing European TDMA cellular standard, the 1.9 GHz GSM system

- One new low-tier TDMA system referred to as the PACS system

- One new Wideband CDMA system

- One composite CDMA/TDMA system

- One low-tier TDMA system based on the existing DECT system

1.4 ORGANIZATION OF THE BOOK

The remainder of this book is organized as follows. Chapter 2 will review the basic elements of the digital wireless communications technology. Chapters 3 and 4 will give in-depth examples of the actual TDMA and CDMA systems using two of the seven standards: the IS-136-based TDMA system and the IS-95-based CDMA system. Finally, Chapter 5 presents highlights of the remaining five standards.

2

DIGITAL WIRELESS
COMMUNICATIONS TECHNOLOGY

In this chapter, we first discuss a definition and the advantages of digital communications systems. We then discuss the main elements of the digital wireless technology.

2.1 WHAT IS A DIGITAL COMMUNICATIONS SYSTEM?

What is a digital communications system? The word *digital* comes from a Latin word meaning *finger*, and, in the present context, has a connotation of the meaning *countable*. A set of discrete elements is called a *countable set*. For example, a set of all integers is a countable set. On the other hand, a set consisting of a continuum of elements is a non-countable set. For example, a set of all real numbers between 0 and 100 is a continuum and is a non-countable set. In a digital communications system, there are two fundamental quantities which are made countable: the *time* at which the information is transmitted, or communicated, and the *magnitude* of the signal, or information, being communicated.

To illustrate this concept, let us start with a discussion of an analog system for a moment, in which both the time and the magnitude are not countable. Consider a graph of the temperature of a room taken continuously over a certain period of time where the abscissa shows the time and the ordinate, the magnitude of the temperature. The graph is continuous on two dimensions: time and magnitude. Even though the time interval under consideration is finite, the number of time points where the temperature is measured is infinite. The time points at which the temperature is taken form a non-countable set. Furthermore, even if the magnitude of the temperature is bounded by a minimum and a maximum, there are still an infinite number of different values of the magnitude in this range: the different values of the magnitude of the temperature measured over the range form a non-countable set.

Suppose now that this temperature information is to be communicated from one end to another through a communications system. At the transmitting end, or the *source*, the temperature is converted to (or represented by) a signal suitable for transmission such as an electrical signal; and, at the receiving end, or the *sink*, the signal is converted back to the temperature. In an analog communications system, the source information is communicated from one end to the other continuously both in time and in magnitude. The electrical signal representing the temperature is transmitted continuously in time and there is an infinite number of possible values of the electrical signal amplitude that may be transmitted over this time. This means that the incidence of communications occurs over a continuum of time, i.e., a non-countable set of time, and the magnitude of information transmitted takes on an infinite number of possible values, i.e., a non-countable set of magnitude.

How then is a digital communications system different from the analog communications system described above? In a digital communications system, the two quantities, *time* and *magnitude (or amplitude)*, are made countable. The process of making the time of communications countable is called *sampling*; and that of making the amplitude countable, *quantization*.

To begin the sampling process, the continuum of time is first discretized to form a countable set of time points. Typically, these discrete points are defined to be the time points nT, where n is an integer and T, the sampling time interval. The amplitude of the signal is sampled and processed for transmission only at these discrete time points, and, therefore, the information is communicated between the source and the sink only at these *countable* points in time.

Several different quantization methods will be discussed in Section 2.4. However, one particular quantization method is briefly discussed here to complete the definition of the digital communications system. In this method, the dynamic range of the signal is subdivided into a finite number of contiguous intervals in a fixed increment. Once these subdivisions of the entire range of possible values are defined, a sampled signal amplitude can be approximated by taking the interval where the sampled value falls and taking the end value of the interval which is closer to the sampled value. With the quantization, there are a *countable* number of possible signal amplitudes that need to be communicated to the other end.

Using our temperature example, suppose that the dynamic range of the analog electrical signal representing the temperature is from 0 to 20 volts. A sampled signal may take on any value in this range. Without the quantization, there are an infinite number of possible amplitudes that need to be transmitted. Suppose that this 20-volt signal range is subdivided into 20 intervals in 1-volt increments, and further suppose that a sampled value is 6.7 volt. This sample of 6.7 volts falls in the interval of 6 – 7 volts and is approximated by 7 volts, taking the end value to which the sampled value is closer. In this method, therefore, there are only 20 possible values that need to be transmitted. The quantization process will be discussed further when we discuss the speech coding.

At this point, the reader may have a doubt about the statement made earlier that, in a digital communications system, the incidence of communications occurs only at discrete points in time. In fact, the following question has been frequently asked:

> *Why is it that certain systems are known as digital wireless communications systems and yet use a continuous microwave analog signal for information transmission?*

Such confusion stems from the fact that the signal carrying the information referred to as the *carrier* is in fact continuous even though the incidence of communications occurs at discrete time points. To dispel this confusion, the concept of modulation needs to be briefly introduced here, which will be discussed further later.

The modulation is a process in which the carrier signal is modified, or *modulated*, according to the information being transmitted. The information is being communicated from the source to the sink by means of the changes in the carrier signal. Whenever such a change occurs, a new piece of information is communicated from the source to the sink; while there is no change in the carrier, no new piece of information is communicated. Because, in a digital communications system, the changes in the carrier signal occur only at discrete time points, the incidence of communications occurs also only at these discrete time points. (In contrast to this, in an analog communications system, the carrier changes continuously according to the continuous source signal.)

2.2   ADVANTAGES   OF   DIGITAL COMMUNICATIONS SYSTEMS

One of the main advantages of the digital communications system is its robust performance in the presence of transmission impairments such as the transmission loss and the noise. The loss attenuates the signal strength, and the noise adds unwanted components to the desired signal. These impairments create bit errors in the digital system and the signal attenuation and distortion in the analog system. In an analog system, repeaters are used to compensate for the signal attenuation. However, the same gain of the repeater that boosts the desired signal strength also boosts the noise level as well and, as a result, the signal-to-noise ratio at the output of the repeater remains the same. This means that, once the noise is introduced into the desired signal, it is physically impossible to make the signal pristine again at the output of the repeater. Because of this, the analog system is sensitive to the noise, and its performance degrades continuously as the noise level increases.

In a digital system, regenerators are used instead of repeaters. Instead of mealy boosting the signal strength, the regenerator determines whether the information-carrying bit is 1 or 0 based on the received signal at the input. Once the decision of 1 or 0 is made, a fresh uncorrupted signal representing that bit is transmitted at the output of the regenerator. Given that the 1/0 decision is correct, the quality of the output signal at the regenerator is made as perfect as that at the source. Because of this, the performance of a digital system can be made to remain relatively constant up to a certain threshold of bit error rate. In a digital system, therefore, the performance is less sensitive to the noise, and it is physically possible to maintain a very high level of steady performance over a certain range of transmission impairments.

The regenerator's ability to detect 1 or 0 correctly depends on the design of the terminal equipment where the speech encoding/decoding and the channel coding are introduced. Therefore, in a digital system, the overall system performance can be controlled by the design of the terminal equipment independent of the length of the system, which is an advantage.

The following is a brief summary of other advantages of the digital system:

■   *Low cost interface to digital switching systems*

■ *Easy signaling.* In a digital system, a new signaling element can be introduced by changing one bit, which is much simpler to implement than analog signaling. Furthermore, these signaling bits are easily multiplexable with the primary signal traffic and, as a result, the channel bandwidth can be used more efficiently.

■ *Easy information storage.* Digital signals are easy to store at the source, the sink and other elements of the communications system.

■ *Easy error-protection.* The advances in the communications theory such as those in the digital signal processing and the mathematical coding theory can be applied to provide error protection for the digital signals.

■ *Easy encryption.* Information in the form of binary bits is easier to deal with for its encryption and the advances in the field can be applied.

■ *Easy multiplexing.* A digital communications system converts different types of source signals to a common binary bit stream, and, therefore, makes the multiplexing of different source signals whether they are speech, data, fax, video, etc. on the same facility easy.

■ *Easy packetization*

As everything else in engineering, however, the choice between the analog and the digital systems depends on the tradeoffs of one benefit with another.

2.3 MAIN TECHNOLOGY ELEMENTS OF DIGITAL WIRELESS COMMUNICATIONS SYSTEMS

Figure 2.1 shows a block diagram of the digital wireless communications system. As shown in the figure, the following four main technology elements characterize a wireless communications system:

■ Source coding

■ Channel coding

■ Modulation

- Physical channel creation

  - Duplexing methods
  - Multiple access methods

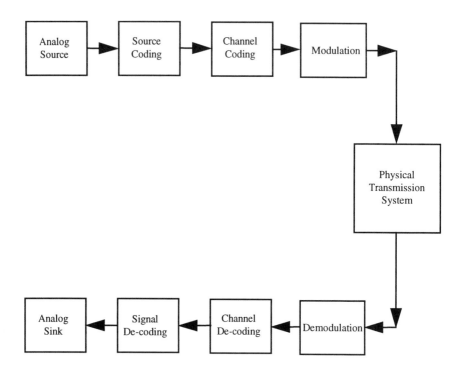

Figure 2.1 A block diagram of the digital wireless communications system

2.4   SOURCE CODING

The source coding deals with the time- and amplitude-discretization of the analog source signal, and, therefore, is tantamount to the definition of the digital communications system discussed in Section 2.1. Typical analog source signals in communications systems are speech and image signals. When the source signal is speech, the subject matter is the speech coding; and, when it is image, the image coding. In this section, we are only concerned with the speech coding. It should also be pointed out that the source signal may already be in a discrete form such as the binary outputs of a computer. Such binary source data are not subject to the source coding and are directly inputted into the channel coding process.

The speech coding techniques are broadly classified into the following two main categories depending upon how the discretization is performed. In both techniques, the time discretization is the same. However, the amplitude-discretization is different in the two techniques.

■   Waveform coding

■   Linear Predictive Coding (LPC)

In practice, these two techniques are sometimes combined to produce hybrid coding techniques.

Subsection 2.4.1 discusses the waveform coding, and Subsection 2.4.2, the Linear Predictive Coding.

2.4.1 The Waveform Coding

In the waveform coding technique, the waveform of the source signal is approximated by mimicking the amplitude-versus-time waveform[1]. In this technique, the speech coding consists of sampling the amplitude of the waveform, approximating the sampled value by the quantization process, and representing, or digitally encoding, the quantized amplitude value by a number of binary bits for transmission. At the sink, the received digital signal is decoded to reconstruct the original waveform as faithfully as possible.

Sampling

Nyquist proved mathematically the following theorem:

> *If the amplitude of a bandlimitted source signal is sampled instantaneously at regular intervals at the rate at least twice the high end frequency of the bandwidth, the samples contain complete information of the source signal so that the original source signal can be perfectly reconstructed from the samples received at the sink, assuming an ideal communications channel which introduces no noise during the transmission of the samples.*

For example, the human speech signal has most of its energy contained in the frequency band of from 0 to 4 kHz – unless of course the speaker happens to be a tenor or soprano and is currently in the middle of his/her performance. In the speech coding, the sampling is based on this 4-kHz speech signal. The minimum sampling rate required to preserve all of the information contained in this 4-kHz band is, applying Nyquist' theorem, 2 x 4 kHz, which is 8000 samples per second.

Nyquist's theorem implies that the sampling process does not incur a loss of information as long as the sampling rate is at least twice the highest component frequency of the signal. In contrast to this, the quantization process to be discussed below is an approximating process, and, by its nature, incur a loss of information, which manifests itself as a noise, or a quantization error.

Quantization

The information-bearing source signal such as the electrical signal produced by the speech acoustic pressure exerted on the telephone mouthpiece has a continuous amplitude. For a digital communications system, this continuous amplitude needs to be discretized. This amplitude-discretization is accomplished by the process known as *quantization*. The quantization process is a process in which an infinite number of possible amplitude values are approximated by a finite number of values.

Figure 2.2 illustrates the sampling, quantization and coding process.

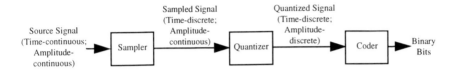

Figure 2.2 The quantization process

Let the continuous-amplitude input to the quantizer be denoted by x and the quantized out be y_i, where $i = 1, 2, \ldots, N$, $y_j > y_i$ for $j > i$. The quantizer maps the sampled amplitude value x to an element y_i of the countable set $\{y_i, y_2, \ldots, y_N\}$ by the following function:

$$y_i = Q(x).$$

To define the function $Q(.)$, the dynamic range of the source signal, x, is subdivided into N intervals and the function $Q(.)$ is a step function as illustrated in Figure 2.3.

Each quantized amplitude is represented by a unique binary bit pattern. Suppose that the number of bits used in this bit-pattern coding of the quantized amplitude is R. Since each binary bit can represent two possible values, 1 or 0, R bits can represent 2^R different values, i.e., the number of quantization levels, N, is given by:

$$N = 2^R.$$

For example, if R = 4, then N = 16, and the 16 quantized levels, $y_1, y_2, \ldots, y_{16}$, may be coded as follows:

Coder Input	Coder Output
y_1	0000
y_2	0001
y_3	0010
.	. . . .
.	. . . .
y_{15}	1110
y_{16}	1111

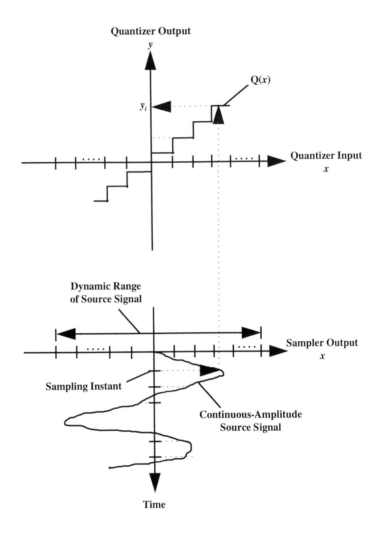

Figure 2.3 The quantizer function Q(.)

Conversely, if the number of quantization levels is to be N, the number of bits required to code one quantized amplitude is given by:

$$R = \log_2 N.$$

The quantization is an approximation process and, therefore, introduces an error referred to as the quantization error. The quantization error is the difference between the actual sampled amplitude, x, and the nearest value assigned to the sample, y_i. Let the sampled value at the sampling instant, nT, be $x(n) = x(nT)$, the corresponding quantized value, $y(n) = y(nT) = Q\{x(nT)\}$. Then the quantization error at the sampling instant nT, $e(n) = e(nT)$, is given by:

$$e(n) = x(n) - y(n).$$

Companding

If the dynamic range of the source signal is subdivided in uniform increments, the quantization is called a uniform quantization; otherwise, nonuniform quantization. One of the drawbacks of the uniform quantization is that the quantization errors produced at the low values of the sampled amplitude are relatively more pronounced than those at the high values. To make the relative effect of the quantization error more uniform, the process known as *companding* (*compressing* and *expanding*) is used. In this method, the input source signal is first compressed by a non linear function of the original source signal, x. The compressed signal, say x' is then used as the input to the uniform quantizer. The output of the uniform quantizer, say y', is expanded to produce y by the inverse function of the compressor function, which is then coded by the binary bit patterns.

There are two commonly used compressor functions: the μ-law companding and the A-law companding as given below:

μ-law

$$x' = c(x) = [\log_e(1 + \mu|x|) / \log_e(1 + \mu)] \operatorname{sgn}(x)$$

where $\operatorname{sgn}(x)$ means the sign of x.

A-law

$$x' = c(x) = [A|x| / (1 + \log_e A)] \operatorname{sgn}(x) \qquad 0 \le |x| \le 1/A$$

$$x' = c(x) = [(1 + \log_e A|x|) / (1 + \log_e A)] \operatorname{sgn}(x) \qquad 1/A < |x| \le 1$$

The μ-law companding is used typically in the North American and Japanese digital systems, and the A-law companding, in the European systems.

The Bit Transmission Rate

Let the sampling rate of the source signal be f_s (samples/second) and the number of bits used to code each sample, R. Then, the bit transmission rate, B (bits/second), of the source is given by the following equation:

$$B = f_s R \text{ (bits/sec)}.$$

For a given dynamic range of the source signal, as more quantization levels are used, the granularity of the step function, $Q(x)$, becomes finer and, as a result, the quantization error is reduced. On the other hand, the more quantization levels are used, the more bits are required to code the quantized amplitudes, and, as a result, the source signal demands a higher bit transmission rate requiring more bandwidth of the communications channel.

For most of the speech coding techniques used in practice, the sampling rate is 8000 samples/second. For example, with the sampling rate of 8000 samples/second, if $R = 8$, the source bit transmission rate is 64 kb/s; if $R = 4$, B = 32 kb/s.

The following major types of waveform coding exist which differ primarily in the quantization method employed:

- Pulse Code Modulation (PCM)

- Differential PCM (DPCM)

- Adaptive DPCM (ADPCM)

- Delta Modulation (DM)

- Sub-band coding

In the next several sub-subsections, these waveform coding techniques are discussed. The PCM, DPCM, ADPCM and DM were originally developed for digital transmission systems such as the T1 lines where the information-carrying signal on the transmission line itself is a train of digital pulses. These same techniques are used here for the source coding but the word *modulation* in the current context should not be confused with the modulation techniques used for transmission to be discussed in Section 2.6.

The Pulse Code Modulation

The Pulse Code Modulation (PCM) is the most bit-consuming speech coding technique. Nevertheless, the PCM is most widely deployed in the telephone network because it is the first telephone speech coding system developed, and, for that reason, the PCM is commonly regarded as the reference system against which other speech coding systems may be calibrated.

The PCM is the simplest, low-complexity speech coding technique. In the PCM, the source speech signal is sampled at 8000 samples per second, and the dynamic range of the speech is quantized into 256 quantization levels so that each sample is approximated by one of these 256 quantized levels. Therefore, in the PCM, eight bits are required to code each sample:

$$R = \log_2 256 = 8 \text{ bits/sample,}$$

and the bit transmission rate of the PCM is 64 kb/s:

$$B = (8000 \text{ samples/sec })(8 \text{ bits/sample}) = 64 \text{ kb/s.}$$

The Differential PCM

The speech signal is a stochastic process with a high degree of autocorrelation. Because of this property, the adjacent samples of the speech signal are highly redundant. Based on the autocorrelation analysis of a number of previous samples, the next sample can be statistically predicted and, at the next sampling instant, the actual sample is taken. As discussed in the previous subsection, in the PCM, the total amplitude of this new sample is quantized, coded and transmitted. In the DPCM, however, this sampled value is first compared with the predicted value and the difference, referred to as the prediction error, is quantized, coded and transmitted. Since the prediction error has a much smaller dynamic range than the absolute signal amplitude itself, a performance gain may be achieved by the DPCM for the same number of quantization levels.

Let:

the, be

$x(n-1), x(n-2), \ldots, x(n-m)$ = previous m samples counted from the current sampling instant, $n = nT$

$x_p(n)$ = predicted value at sampling instant n

$x(n)$ = actual sampled value at n

$e_p(n)$ = prediction error at n

Then, the prediction, $x_p(n)$, is given by a linear combination of the m previous samples as follows:

$$x_p(n) = \alpha_1 x(n-1) + \alpha_2 x(n-2) + \ldots \ldots +. \alpha_m x(n-m)$$

$$e_p(n) = x(n) - x_p(n)$$

where

$$\alpha_1, \alpha_2, \ldots \ldots, \alpha_m = \text{a set of the predictor coefficients.}$$

The prediction error, $e_p(n)$, is quantized, coded and transmitted, by the same quantization process discussed for the PCM.

There are two variants of the DPCM: one is the Adaptive DPCM (ADPCM) and the other, Delta Modulation (DM). These two related speech coding techniques are discussed in the next two subsections.

The Adaptive DPCM

The ADPCM is same as the DPCM except that, unlike in the DPCM where the predictor coefficients are constants, in the ADPCM, the predictor coefficients are updated at each sampling instant and therefore are a function of time as follows:

$$x_p(n) = \alpha_1(n)x(n-1) + \alpha_2(n)x(n-2) + \ldots \ldots +. \alpha_m(n)x(n-m).$$

$$e_p(n) = x(n) - x_p(n)$$

In a typical ADPCM, the sampling rate of the source signal is 8000 samples per second. and the quantized value of the prediction error, $e_p(n)$, is coded using four bits, i.e., $R = 4$. Therefore, the number of quantization levels is:

$$N = 2^4 = 16.$$

Compare this with the number of quantization levels, N, for the PCM, which is 256. The N for the ADPCM is only one sixteenth of that for the PCM.

The bit transmission rate of the source signal encoded by the ADPCM is:

$$B = (8000 \text{ samples/sec})(4 \text{ bits/sample}) = 32 \text{ kb/s}.$$

The ADPCM provides the landline "toll quality" speech performance, which will be discussed in Subsection 2.4.3.

The Delta Modulation

The Delta Modulation (DM) is an extreme case of the DPCM and may be regarded as the one-bit DPCM, where a single bit is used to code the signal at each sampling instant. The basic concept is as follows. The source signal is approximated by a staircase function with a fixed step size, $\pm \Delta$. At each sampling instant, the sign of this step change is determined based on the difference between the sample at that sampling instant and the latest step-function approximation of the signal.

Let:

$$x(n) \quad = \quad \text{sampled value of } x \text{ at } n = nT$$

$$x_s(n-1) \quad = \quad \text{step-function approximation of } x \text{ as of } (n-1)$$

$$x_s(n) \quad = \quad \text{step-function approximation at n,}$$

Then $x_s(n)$ is determined as follows:

$$x_s(n) = x_s(n-1) + \Delta \qquad \text{if } \{x(n) - x_s(n-1)\} \text{ is positive}$$

$$x_s(n) = x_s(n-1) - \Delta \qquad \text{if } \{x(n) - x_s(n-1)\} \text{ is negative.}$$

In the DM, therefore, only the sign of the quantity $\{x(n) - x_s(n-1)\}$ needs to be coded, which takes one bit: 1 for the positive difference, and 0 for the negative difference.

Since only one bit is used to code the signal at each sampling instant, the bit transmission rate of the source signal is same as the sampling rate:

$$B = f_s \text{ (bits/sec)}.$$

For example, in the DM, if the speech signal sampling rate is 8000 samples/sec, the source bit transmission rate would be 8 kb/s.

The Sub-band Coding

In all of the four waveform coding techniques discussed above, the source waveform used as the input to the coding process is treated as a single signal with its full bandwidth. In the sub-band coding technique, the input signal is first decomposed into a number of its frequency components, and each of these frequency components is encoded separately. Since, in this technique, the encoding is a function of frequency as well as time, the sub-band technique involves both time-domain and frequency-domain analyses.

One major advantage of the sub-band coding is that it provides an additional flexibility for the engineering tradeoff. Since, in this technique, the individual frequency components are treated separately, the number of bits assigned to code the different frequency components can be made variable allowing the designer to use more bits for more important frequency components, and less, for the less important components.

The concept of the sub-band coding is fairly simple as follows:

1. The total spectrum of the source speech signal is decomposed into L sub-band frequency components. This is accomplished by using L bandpass filters, BPF_1, ..., BPF_L, with bandwidth ΔW_1, ..., ΔW_L, respectively, placed in parallel with the source signal as the common input. If the total bandwidth is W and the sub-bands are created to have an equal width, $\Delta W_k = W/L$, for k = 1, 2, . . . , L. The outputs of the bandpass filters are the L sub-band components of the original signal.

2. These L sub-band signals are sampled.

3. At each sampling instant, the L sampled values are quantized and coded separately by L encoders in parallel. The outputs of these L encoders are L separate binary bit streams. These L bit streams are multiplexed and transmitted to the other end, i.e., the sink.

4. At the sink, the reverse process of Steps 1 and 2 takes place: the total received bit stream is de-multiplexed into L separate bit streams; the L bit streams are

decoded by L decoders in parallel; and the individual frequency sub-band components are superimposed, i.e., summed, to reconstruct the source signal.

2.4.2 The Linear Predictive Coding

In the Linear Predictive Coding (LPC) technique, the human vocal tract is modeled and the parameters of the model are estimated based on the individual speech signal. These parameters are coded and transmitted to the sink. The sink employs a replica of the vocal tract model used at the source and synthesizes the speech signal using the model parameters sent from the source. The amplitude-discretization is realized by representing the continuous speech signal with a handful of model parameters.

The following is a list of different types of the LPC techniques:

- Residual Excited Linear Predictive (RELP) coding

- Code Excited Linear Predictive (CELP) coding

- Vector Sum Excited Linear Predictive (VSELP) coding

- Regular Pulse Excitation-Long Term Prediction (RPE-LTP) coding

With these LPC coding techniques, the source bit transmission rate can be made as low as that in the range of 1 – 15 kb/s.

2.4.3 The Performance of the Speech Codec

The Mean Opinion Score

The most commonly-used measure of the speech coder/decoder (codec) performance is the subjective evaluation index referred to as the Mean Opinion Score (MOS). The MOS is determined for a given speech codec as follows. Test case speech samples are prepared by reading "phonetically balanced" sentences into the codec and recording the speech output of the codec into a tape. The recorded speech represents the speech that has undergone the speech encoding and decoding process. The same speech samples are played to a group of subjects individually in a laboratory setting. After listening to the speech samples, the subjects are then asked to rate the speech quality on a five-point rating scale consisting of *excellent, good, fair, poor,* and *unsatisfactory*.   Each of these five ratings is assigned a numerical score as follows: 5 for *excellent,* 4 for *good,* 3 for *fair,* 2 for *poor,* and 1 for *unsatisfactory*. The subject's responses are tallied and the MOS is calculated as the average score of the ratings as follows:

$$MOS = (5N_e + 4N_g + 3N_f + 2N_p + 1N_u) / N$$

where

N_e, N_g, N_f, N_p, N_u = Numbers of *excellent, good, fair, poor, unsatisfactory*

N =   Total number of ratings   = $N_e + N_g + N_f + N_p + N_u$

As an example, suppose that test speech samples are played to 100 subjects, and the number of their ratings are as follows:

Rating Category	Number of Subjects
Excellent	10
Good	20
Fair	40
Poor	20
Unsatisfactory	10
Total	100

The MOS of the codec is calculated as follows:

$$MOS = \{(5)(10) + (4)(20) + (3)(40) + (2)(20) + (1)(10)\}/100 = 3.0$$

The %Good-or-Better and the %Poor-or-Worse

Although the MOS is a simple indicator of the average, or mean, performance, it is not a good measure of the distribution of the performance. If the variability of the subjective opinion is large, there could be a significant number of subjects who are unhappy about the performance even though the MOS might seem reasonable. To provide the information about the distribution of the subjective opinions, in addition to the MOS, two other subjective measures are commonly-used which are derived from the same subjective testing results. They are the *%Good-or-Better* and the *%Poor-or-Worse*.

The *%Good-or-Better* is the percentage of the subjects who rate the speech performance as either *good* or *excellent* (i.e., *better than good*); and the *%Poor-or-Worse* is the percentage of the subjects who rate the speech performance as either *poor* or *unsatisfactory* (i.e., *worse than poor*). The intent of the *%Good-or-Better* is to keep the percentage of the population experiencing a good or better performance at a certain high level regardless of the total distribution; similarly, the intent of *%Poor-or-Worse* is to keep the percentage of the population experiencing a poor or worse performance at a certain low level regardless of the total distribution. In the example above, *%Good-or-Better* is 30% and the *%Poor-or-Worse*, 30%.

System Performance

The so-called "toll quality" speech performance is commonly regarded as the MOS of 4.0. The speech performance of the analog cellular system such as the Advanced Mobile Phone Service (AMPS) is generally regarded to be poorer than the toll quality. This is understandable but the poorer speech performance is not due to speech coding because the analog system does not of course use the speech coding techniques discussed above. However, it often serves as a reference against which the digital wireless system performance is compared.

For digital wireless communications systems, the ADPCM speech codec is generally known to have the best speech quality. This is of course only fair because it uses 32 kb/s of bandwidth, which is the highest among the speech codecs used in the digital wireless systems today.

The LPC techniques have dramatically reduced the bit transmission rate, i.e., the bandwidth requirements, using a completely different approach of human vocal tract modeling and speech synthesis. The synthesized speech, however, is generally known to sound artificial to the extent that the speaker's own family member at times does not recognize the voice on the phone.

2.5 CHANNEL CODING

2.5.2 Error Control Techniques

The output of the speech encoder is further digitally processed and encoded for the error control during the transmission through the communications channel. This process of coding is referred to as the channel coding as opposed to the source coding.

Fundamentally, the error control over the communications channel involves providing error detection and error correction capabilities. The following are examples of commonly used error detection and error correction techniques:

The error detection techniques include:

■ Cyclic redundancy check (CRC)

The error correction techniques include

■ Block coding

■ Convolutional coding

Another error control operation, called *interleaving*, provides neither an error detection capability nor an error correction capability but it indirectly enhances the codec's error correction capability.

2.5.2 The Cyclic Redundancy Check

The general operation of the Cyclic Redundancy Check (CRC) is described. Given a k-bit block of binary data, an n-bit CRC sequence is generated, which is appended to the k-bit data block. The n-bit CRC is determined in such a way that the resulting (k + n) bit sequence is exactly divisible by some predetermined bit string which is (n + 1) bits long. At the receiver, this division operation is performed and, if there is no remainder, the received data is assumed to be error-free; otherwise, errored thereby allowing error detection.

The process of obtaining the CRC bit sequence is explained below.

1. Let the k-bit information-carrying data string be $m_k m_{k-1} m_{k-2} \ldots m_1$ and (n + 1)-bit divisor bit pattern be $d_n d_{n-1} d_{n-2} \ldots d_1$. Using these bits as coefficients, the following two polynomials are formed:

$$M(X) = m_k X^k + m_{k-1} X^{k-1} + m_{k-2} X^{k-2} + \ldots m_1 X^1$$

$$D(X) = d_n X^n + d_{n-1} X^{n-1} + d_{n-2} X^{n-2} + \ldots d_1 X^1$$

2. Perform the following division operation:

$$X^n M(X)/D(X) = Q(X) + \{R(X)/D(X)\}$$

3. Discard the quotient Q(X). The coefficients of the remainder polynomial R(X), $r_n r_{n-1} r_{n-2} \ldots r_1$, form the CRC bit sequence.

4. The (k + n)-bit sequence to be transmitted is $T(X) = X^n M(X) + R(X)$, which yields:

$$m_k m_{k-1} m_{k-2} \ldots m_1 \, r_n r_{n-1} r_{n-2} \ldots r_1,$$

The following are widely used divisor polynomials:

- CRC-12: $D(X) = X^{12} + X^{11} + X^3 + X^2 +. X + 1$

- CRC-16: $D(X) = X^{16} + X^{15} + X^2 + 1$

- CRC-CCITT: $D(X) = X^{16} + X^{12} + X^5 + 1$

- CRC-32: $D(X) = X^{32} + X^{26} + X^{23} + X^{22} + X^{16} + X^{12} + X^{11} + X^{10} + X^8 + X^7 + X^5 + X^4 + X^2 +. X + 1$

Example

Let the message to be transmitted is M = 1010001101 (k = 10 bits) and the divisor pattern D = 110101 (n = 6 bits). The CRC bits or Frame Check Sequence (FCS) to be calculated is five bits as follows. Using modulo 2 arithmetic, the message sequence M is first multiplied by 2^5 (i.e., X = 2) to yield 101000110100000. This is then divided by the divisor pattern D by modulo 2 arithmetic to yield the quotient Q = 1101010110 and the remainder R = 1110. The (k+n)-bit sequence transmitted is by $T(X) = X^n M(X) + R(X)$ or $T = 2^5 M + R$, 101000110101101which is obtained as follows:

$$1\ 0\ 1\ 0\ 0\ 0\ 1\ 1\ 0\ 1\ 0\ 0\ 0\ 0\ 0$$

$$+\qquad\qquad\qquad\qquad 1\ 1\ 0\ 1$$

$$\overline{1\ 0\ 1\ 0\ 0\ 0\ 1\ 1\ 0\ 1\ 0\ 1\ 1\ 0\ 1}$$

2.5.3 The Block Coding

The data is grouped into k-bit blocks. To each of the k bits for a block, (n-k) redundant bits called parity bits are appended. The resulting n-bit sequence for transmission is called (n, k) block code. With a (n, k) block code, for every n bits transmitted, only k bits are information-bearing. The code rate is defined as:

$$k/n.$$

At the sink, the received bit string is decoded by calculating the Hamming distances. For a (n, k) block code, 2^k Hamming distances must be computed to determine the smallest Hamming distance and the corresponding information bits.

The concept of the Hamming distance is explained by the example of $k = 2$ and $n = 5$, i.e., (n-k) = 3 parity bits. Suppose that a certain (5, 2) block code algorithm generates the following parity bits to be appended to the two information bits:

Information Bits	Parity Bits
00	000
01	011
10	100
11	111

Suppose that the received bits are 10111, and the decoder must detect the two information bits. The decoder knows the coding rule. Since $k = 2$, four Hamming distances must be calculated, i.e., one for each of the four possible five-bit strings listed above: 00000, 01011, 10100, and 11111. The received bit pattern 10111 is compared bit by bit with each of the above four strings and the number of times the two compared bits are different is counted. The total number of times they are different is the Hamming distance. In our example, the four Hamming distances are:

4, 3, 2, and 1. Taking the smallest Hamming distance, which is 1, the decoder assumes that the transmitted bits are 11.

2.5.4 Interleaving

The bit errors due to the random noise such as the thermal noise of the circuit, tend to occur as isolated errors scattered in time with a random inter-arrival time. These single isolated errors are easily detectable and correctable by the error detection/correction techniques discussed above. In a wireless communications system, an important cause of the bit error is the fading. When the system experiences a fading, a burst of bit errors may occur during the fading period. When bit errors occur consecutively in bursts, the effectiveness of the error detection/correction schemes is severely reduced. Therefore, even though the total number of bit errors over a certain longer interval of time is kept the same so that both bit error patterns have the same average bit error rate, the bursty bit error pattern is much harder to deal with by the error detection/correction schemes than the randomly scattered bit error pattern.

The interleaving is the process by which the natural transmission order of the bit sequence is deliberately altered at the transmitting end in such a way that the burst of bits in error which are consecutive in the altered order are in fact separated from each other when the bit sequence is "de-interleaved" to its original natural order. When a fading hits a stream of interleaved bits, a burst of bit errors may occur consecutively. However, these errored bits are in fact separated in the natural order after the de-interleaving.

Figure 2.4 illustrates the effect of interleaving. The natural transmission order of bits numbered 1, 2, 3, 4, 5, 6, 7, 8, 9, 10 are interleaved and transmitted in the following order: 1, 6, 2, 7, 3, 8, 4, 9, 5, 10. During transmission, two consecutive errors occur on bits 7 and 3 shown by X's. At the receiver, the de-interleaver puts the bits in the original order. When the bits are rearranged in the original order, the two consecutive errors on bits 7 and 3 are separated. These separated errors are easier to handle by the regular error-correction techniques such as the CRC than the clustered errors.

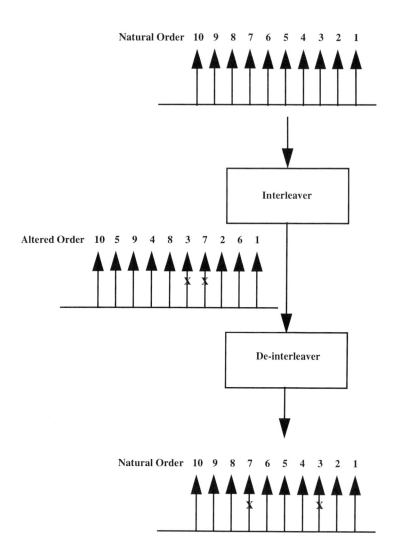

Figure 2.4 The effect of interleaving

2.6 MODULATION

2.6.1 The Basic Modulation Techniques

The binary information bits are typically represented by some form of electrical pulses. However, these pulses are not suitable for transmission over the air. In the wireless communications system, an analog sinusoidal signal referred to as a carrier is used to "carry" the information. The modulation is the process in which the carrier signal is modified, or *modulated*, to reflect the information bits so that, at the receiving end, the information bits are recovered by de-modulating the analog signal.

A sinusoidal carrier may be represented as follows:

$$A\sin\{\omega t + \phi\}$$

where A is the amplitude, $\omega = 2\pi f$, the angular frequency, and ϕ, the phase. There are three main types of modulation techniques depending on which of the three quantities in the carrier is modified: the Amplitude Modulation (AM), the Frequency Modulation (FM) and the Phase Modulation (PM).

The Amplitude Modulation (AM)

Let the information bit at sampling instant n be b(n). The amplitude A is a function of b(n): $A(n) = A(nT) = g\{b(n)\}$. The simplest AM technique is to change the amplitude between two different values keeping the rest of the parameters of the carrier signal the same: A_1 for b(n) = 0 and A_2 for b(n) = 1. The modulated carriers are shown below:

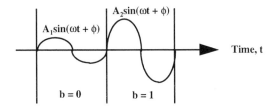

The Frequency Modulation (FM)

In this modulation technique, the frequency, f, is varied according to the information bits: $f(t) = g\{b(t)\}$, where $t = nT$. A simple example is the Frequency Shift Keying where two different frequencies are used: f_1 for $b(t) = 0$ and f_2 for $b(t) = 1$. The modulated carriers are shown below:

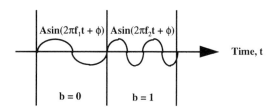

The Phase Modulation (PM)

Similarly, in this case, the phase ϕ is varied: $\phi(t) = g\{b(t)\}$, where $t = nT$. A simple example is the Phase Shift Keying where two different phases are used: ϕ_1 for $b(t) = 1$ and ϕ_2 for $b(t) = 0$. The modulated carriers are shown below:

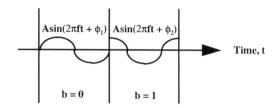

2.6.2 The Symbol Rate versus the Bit Rate

If four amplitudes are used in the AM, each modulated carrier can represent two bits. Similarly, with the FM and the PM, if four different frequencies and four different phases are used, two bits can be represented by one modulated carrier signal by the FM and the PM respectively. The different modulated carriers are called symbols, or equivalently, the groups of bits represented by the modulated carriers are called symbols.

Symbol	Bits	AM	FM	PM
Symbol 1	00	$A_1\sin(\omega t+\phi)$	$A\sin(\omega_1 t+\phi)$	$A\sin(\omega t+\phi_1)$
Symbol 2	01	$A_2\sin(\omega t+\phi)$	$A\sin(\omega_2 t+\phi)$	$A\sin(\omega t+\phi_2)$
Symbol 3	10	$A_3\sin(\omega t+\phi)$	$A\sin(\omega_3 t+\phi)$	$A\sin(\omega t+\phi_3)$
Symbol 4	11	$A_4\sin(\omega t+\phi)$	$A\sin(\omega_4 t+\phi)$	$A\sin(\omega t+\phi_4)$

In the above examples where four different modulated carriers are used, four different symbols are employed; and each of the four symbols represent two bits as follows: Symbol 1 is 00, Symbol 2, 01, Symbol 3, 10 and Symbol 4, 11. The association between the symbol number and the bit pattern is arbitrary. For example, bits 10 may have been named Symbol 2 as well.

Let the number of symbols be S and the number of bits represented by each symbol, B. Then the following relationship holds true:

$$S = 2^B.$$

In other words, to represent B bits per one incidence of carrier change, $S = 2^B$ different modulated carriers must be employed.

Conversely, given S, B is given by:

$$B = \log_2 S.$$

The rate at which the symbols change, i.e., the rate at which the carrier changes, is referred to as the symbol rate, and the unit is the baud. One baud is one symbol per second. Clearly, if only two symbols are used, then S = 2 and B = 1 and each symbol represents one bit. Therefore, for S = 2, the bit transmission rate is equal to the symbol rate, i.e., n kb/s bit transmission rate results in n kbaud of symbol transmission rate. In general, for a given number of symbols, S, the bit rate, n kb/s, and the symbol rate, m kilo-bauds, are related by the following equation:

$$n = m \log_2 S \text{ kb/s.}$$

For example, a transmission system with a modulation scheme with four symbols (S = 4) operating at 9.6 kbauds are transmitting the bits at the rate of 19.2 kb/s as follows:

$$n = 9.6 \log_2 4 \text{ kb/s} = 19.2 \text{ kb/s.}$$

2.6.2 The Quadrature Phase Shift Keying

In the Quadrature Phase Shift Keying (QPSK) modulation scheme, four different phase angles are used, thereby creating four symbols: $\pi/4$, $3\pi/4$, $5\pi/4$, and $7\pi/4$. The amplitude is constant. The QPSK corresponds to the PM example with S = 4 given above. The modulated carriers with these four phase angles and the constant amplitude are expressed as phasors or vectors as shown below:

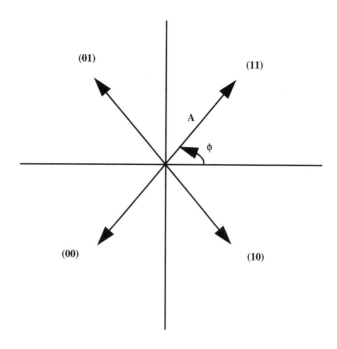

The symbol-to-bit pattern mapping is such that as the phase angle changes from the neighboring angle the two-bit pattern makes one bit change. For example, the two-bit pattern for $\phi = 45°$ is 11 and that for $\phi = 135°$ is 01. Comparing 11 and 01, note that there is one bit difference between the two patterns.

This type of symbol-to-bit pattern mapping is referred to as the Gray coding. The reason for using the Gray coding is as follows. The decoding error is more likely to occur between the adjacent phase angles, i.e., if one phase angle is erroneously decoded, it is more likely to be decoded as the angle next to the true angle. Therefore, by using the Gray coding, only one-bit error is made, instead of both bits, when the true angle is mis-decoded as an adjacent angle. Both bits will be in error if the true angle is mis-decoded as the angle two steps away, which is much less likely.

Since four symbols are used in the modulation each symbol representing two bits, the bit transmission rate is twice the symbol rate.

2.6.3 The π/4-Differential Quadrature Phase Shift Keying

In the differentially-encoded modulation, information is transmitted by the amount of the difference in phase rather than absolute phases. In this method, four values of phase changes, $-3\pi/4$, $3\pi/4$, $\pi/4$ and $-\pi/4$, are used to modulate the carrier. Since four different phase changes are used, each phase change can symbolize two bits. Using the Gray code, the symbol and phase change mapping is shown below:

Odd-Numbered Bit	Even-Numbered Bit	Phase Difference
1	1	$-3\pi/4$
0	1	$3\pi/4$
0	0	$\pi/4$
1	0	$-\pi/4$

The π/4-Differential Quadrature Phase Shift Keying (π/4-DQPSK) has been adopted as the U.S. and the Japanese standards.

38

2.6.4 The Gaussian Minimum Shift Keying

Unlike the QPSK and the π/4-DQPSK, which are both phase modulation (PM) schemes, the Gaussian Minimum Shift Keying (GMSK) is an FM, which uses two symbols, S = 2: each symbol represents one bit. Therefore, in the GMSK, the bit rate and the symbol rate are the same.

2.7 METHODS OF CREATING PHYSICAL CHANNELS

Once the source signal is sampled, quantized, source-coded, and channel-coded and the resulting bits are used to modulate the carrier signal, the modulated carrier must now be transmitted to the other end. In the case of the wireless access system, physical channels must be created over the air to transmit this modulated signal from the Mobile Station (MS) to the Base Station (BS) and *vice versa*. In this section, various methods of creating the physical channels over the air are discussed.

2.7.1 The Finite Resource: Equivalence in the Frequency Domain and the Time Domain

The Finite Resource
in the Frequency-Domain

If roads are to be created in a region, the finite resource would the land available in that region. In communications systems, a commonly used quantity of expressing the finite resource is the bandwidth of the frequency spectrum of a physical medium. For our discussion, the physical medium is the air. The frequency spectrum of the air is a natural resource managed by the FCC. Supposed that a certain amount of bandwidth is allocated by the FCC to be used for a wireless communications system. The physical channels for the system must be created from this limited bandwidth. Denote this total bandwidth available for the system by W Hz.

The Finite Resource
in the Time-Domain

In order to understand the different methods of creating physical channels to be discussed in later sections, we need to understand the equivalence of the finite resource in the frequency-domain and the time-domain. Because of this correspondence between the same finite resource expressed in two different dimensions, a bit transmission rate – which is a time-domain quantity – is often referred to simply as a bandwidth meaning that transmitting the information at the given bit rate consumes a certain amount of bandwidth.

Let us carry this concept further to make a concrete correspondence between the finite resource in the frequency domain and that in the time domain. Suppose that we are given a frequency spectrum with a fixed bandwidth W Hz as the finite resource from which the physical channels must be created. Further, consider a digital bit stream to be transmitted at a bit rate, B b/s. Depending upon the modulation scheme, the corresponding symbol rate, S baud, is determined according to the formula discussed earlier. To transmit the bits at a faster rate, the symbol rate must also be increased accordingly. Therefore, there is a one-to-one correspondence between the bit rate and the symbol rate. For simplicity, in our discussion, let us use the bit rate as the time domain quantity instead of the symbol rate.

Suppose that a binary bit stream is used to modulate a carrier according to a certain modulation scheme and the modulated signal is applied to the input to a signal spectrum analyzer. The scope of the spectrum analyzer will then show the frequency spectrum of the digital bit stream after the modulation. To make the experiment more realistic, the impairments representing the transmission medium may be inserted in the path as well. Let the bandwidth of the digitally modulated signal at the bit rate B be W_b. As the bit rate B is increased, the scope will show an increase in W_b. Increase the bit rate B in this fashion until the bandwidth of the bit stream completely fills the given fixed bandwidth, W. Let this bit rate be B_{max}.

B_{max} is a function of the channel coding, the modulation scheme and the transmission impairments. (The speech coding is excluded assuming that the transmission system begins after the source coding, and defining B_{max} as the maximum bit rate that can be applied to the system by the output of the speech coding.) Engineers have taken pains to develop various ingenious schemes for the channel coding and modulation schemes for the purpose of increasing B_{max} at which to "push the bits through" a given bandwidth. After all of this effort, however, there is still a limit, B_{max}, for a given fixed bandwidth W. This maximum possible bit rate B_{max} is the same finite resource W expressed differently in the time-domain:

frequency-domain finite resource		time-domain finite resource
W Hz	$\Leftrightarrow$	B_{max} b/s

To further quantify the concept that the fixed bandwidth of W Hz can support the bit transmission rate of B_{max} b/s, consider the time axis and mark it with points referred to as *bit positions* at the rate of B_{max}. Since there are B_{max} bit positions in one second, the time distance between two consecutive bit positions is $1/B_{max}$ second. Since the maximum bit rate that can be supported by the bandwidth W is B_{max} b/s, $1/B_{max}$ second is the minimum possible distance between two consecutive bit positions for the system. As will be discussed later, a number of bit positions can be grouped into a *time slot* and a number of time slots, into a *frame*. The time slot length and the frame length are design parameters, and are selected as part of the system design. However, the minimum inter-bit position time cannot be made any smaller than $1/B_{max}$ by definition.

Figure 2.5 illustrates the equivalence of the finite resource in the frequency- and the time-domain in terms of the total bandwidth W and the corresponding B_{max} bit positions per second.

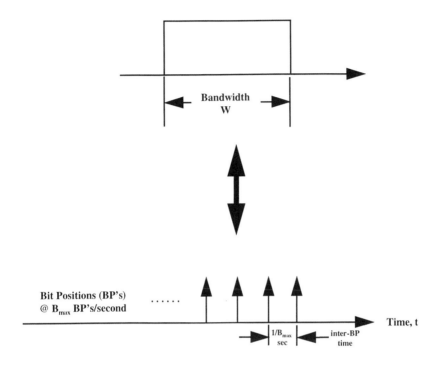

Figure 2.5 The finite resource in the frequency- and the time-domain

2.7.2 Creating Physical Channels: Subdividing the Finite Resource

Creating the physical channels for a communications system amounts to subdividing the finite resource available for the system.

The Types of Physical Channels

Communications involves two ends: transmit end, or the source, and the receive end, or the sink. For the wireless communications systems, the Mobile Station (MS) is at one end, and the Base Station (BS), at the other end, with the wireless link in-between. Three types of communications channel may be considered: full-duplex, half-duplex and simplex channels.

■ *Full-duplex channel.* A full-duplex channel is a communications channel which provides a communications path simultaneously in both directions. This may be compared with a two lane bridge where cars can travel in both directions simultaneously.

■ *Half-duplex channel.* A half-duplex channel is a communications channel which provides a communications path in both directions but not simultaneously. This may be compared with a single lane bridge where cars can travel in both directions alternately.   As will be discussed later, in certain transmission systems, e.g., a Time Division Duplexing system, information bits are transmitted in both directions alternately in time so that the transmission operation resembles that of a half-duplex channel. However, as long as the user does not perceive this alternation, the system may be considered as a full duplex system.

■ *Simplex channel.* A simplex channel is a communications path which provides a communications path in one direction only. This may be compared with a one-way bridge where cars can travel in only one direction.

The Methods of Subdividing Finite Resource

The method of creating physical channels involves allocating the finite resource in two steps: first the finite resource is allocated for the two directions of transmission, and next, the portion allocated for each direction is used to create multiple channels for that direction. These two steps of subdividing the finite resource are referred to as:

■ Duplexing methods

■ Multiple access methods

2.7.2 The Duplexing Methods

There are two main types of duplexing methods:

- Frequency Division Duplexing (FDD)

- Time Division Duplexing (TDD)

The Frequency Division Duplexing

In wireless communications systems, the direction of from the BS to the MS is referred to as the *forward* direction, and that of from the MS to the BS, the *reverse* direction. In the Frequency Division Duplexing (FDD) method, the total available bandwidth W is first allocated for the forward and reverse directions of transmission. Let the total bandwidths allocated for the forward and reverse directions be W_F and W_R, respectively. This allocation is illustrated below:

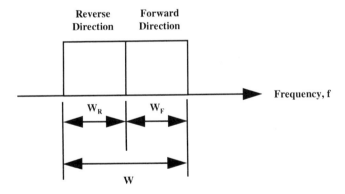

In theory, the two sub-bands W_R and W_F do not necessarily have to be two complete contiguous bands as long as all of the bands allocated to the two directions add up to W_R and W_F, respectively. An example of dividing the total band into noncontiguous sub-bands is illustrated below:

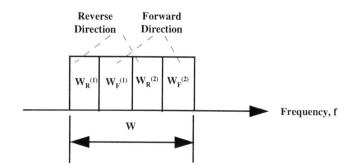

As will be seen later, however, in practice, the contiguous division is more commonly used because it simplifies the frequency administration. One might argue that the contiguous division maximizes the frequency separation between the reverse and forward directions but the same can be accomplished by the noncontiguous division except that the frequency administration would then become more complicated.

The Time Division Duplexing

In the Time Division Duplexing (TDD), the two directions of transmission are created by alternating the transmission in time. The maximum bit rate B_{max} supportable for a given bandwidth W is allocated between the two directions. If this bit rate is equally divided between the two directions, each direction of transmission can support at most $B_{max}/2$. As will be discussed in the TDD/TDMA, a typical method of dividing the resource between the two directions is to have each full duplex channel alternate the transmission in time. This method is also colloquially referred to as a "Ping-Pong" method.

2.7.2 The Multiple Access Methods

The multiple access methods refer to the method of creating multiple channels for each transmission direction. There are three main types multiple access methods:

■ Frequency Division Multiple Access (FDMA)

■ Time Division Multiple Access (TDMA)

■ Code Division Multiple Access (CDMA)

Since the multiple access method depends on the duplexing method, it must be discussed for a given duplexing method. For example, the TDMA is possible for both the FDD and the TDD. Both the FDD/TDMA and the TDD/TDMA will be discussed. However, the FDMA is possible in practice only for the FDD. Therefore, to discuss, the FDMA, the FDD method of duplexing will be assumed. For the CDMA, both the FDD and the TDD are possible in theory. In practice, however, the TDD/CDMA combination is not used yet; only the FDD/CDMA combination exists.

Although we discuss the FDMA, the TDMA and the CDMA separately, the real systems are usually a combination of multiple methods with a possible exception of the FDMA system. For example, a system which may be commonly-referred to as the TDMA system is in fact most likely a combination of the FDMA and the TDMA; and similarly, with a CDMA system.

The Frequency Division Multiple Access

We discuss the Frequency Division Multiple Access (FDMA) system with the FDD where the total bandwidth W is subdivided into W_R and W_F contiguously for the reverse and forward directions of transmission. In the FDMA method, each of the two sub-bands, W_R and W_B, is further subdivided into N smaller bands to create N channels. These smaller bands of frequency are denoted as $W_R^{(1)}$, $W_R^{(2)}$, ..., $W_R^{(j)}$, ..., $W_R^{(N)}$ and $W_F^{(1)}$, $W_F^{(2)}$, ..., $W_F^{(j)}$, ..., $W_F^{(N)}$, respectively. Each pair of the smaller bands, $W_R^{(j)}$ and $W_F^{(j)}$, is a full-duplex channel. The following figure illustrates the FDD/FDMA.

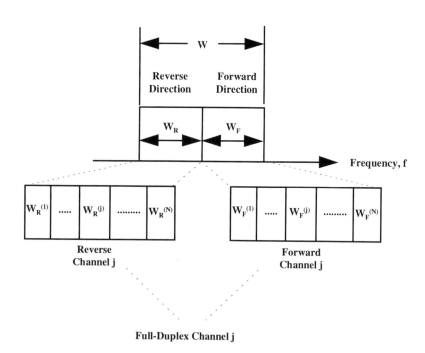

Once the individual physical channels are created by the FDD/FDMA method, the maximum allowable bit transmission rate, $B_{R-max}^{(j)}$ and $B_{F-max}^{(j)}$, of each individual channel is related to the maximum bit rate of the air interface by:

$$B_{R-max}^{(j)} = B_{F-max}^{(j)} = (B_{max}/2)(\ 1/N)\ b/s.$$

The Time Division
Multiple Access

The Time Division Multiple Access (TDMA) is used in combination with both the FDD and the TDD. Both the FDD/TDMA and the TDD/TDMA are discussed.

The FDD/TDMA

In this combination, the FDD part is same as in the FDD/FDMA discussed above: the total bandwidth W is divided into W_R and W_F for the reverse and forward directions, respectively. Each of these two sub-bands is then used independently to create multiple channels using the TDMA technique.

Recall that the total bandwidth W Hz supports the maximum bit rate of B_{max} b/s. Denoting the maximum bit rates supportable on the reverse and forward links with the respective bandwidths, W_R and W_F, by B_{R-max} and B_{F-max}, respectively, we have

$$B_{R-max} = B_{F-max} = B_{max} / 2.$$

The FDD/TDMA process is illustrated in Figure 2.6.

These bit positions are organized into individual channels as follows. A number of consecutive bit positions are grouped into a time interval which is longer than the basic inter-BP time, called the time slot. Time slots are the basic building blocks of individual physical channels. Time slots are grouped into frames so that the same time slot is repeated once in each frame. Let:

Number of bit positions (BP's) in a time slot be	=	N_{BP}
Number of time slots in a frame	=	N_{TS}
Inter-BP time in each direction	=	t_{R-BP} or t_{F-BP}
	=	$2/B_{max}$ second

Then, the time slot duration, t_{TS}, and the frame duration, t_F, are given by:

Time slot duration	=	t_{TS}
	=	$(N_{BP})(t_{R-BP})$
	=	$(2N_{BP})/B_{max}$ second
Frame duration	=	t_F
	=	$(N_{TS})(t_{TS})$

$$= \quad (2N_{TS}N_{BP})/(B_{max}) \text{ sec}$$

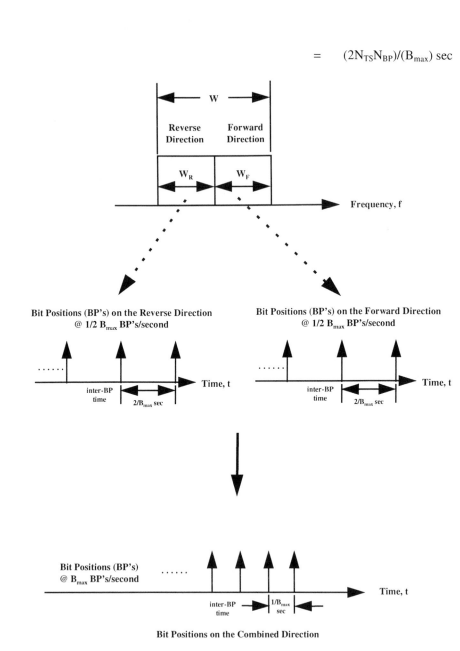

Figure 2.6 The FDD/TDMA process

The relationship between the bit positions, the time slots and the frames is illustrated by the following figure:

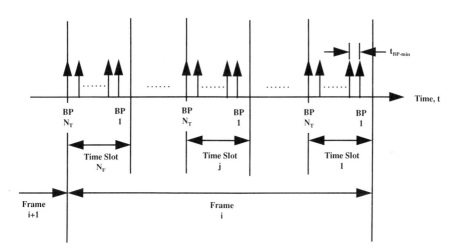

Given the relationship between the bit positions, the time slots and the frames, the individual physical channels in a typical TDMA system are created as follows. Each physical channel is assigned one or more time slots per frame. For the simplicity of illustration, let us suppose that each physical channel is assigned one time slot per frame. Also for the simplicity of illustration, let us consider one direction of transmission at a time, say, the reverse direction from the MS to the BS. Consider the reverse channel, Channel j, which is assigned Time Slot j. Time Slot j occurs once in every frame at the same relative position within the frame. The MS on Channel j is allowed to transmit only during Time Slot j: in each frame, the MS transmits during Time Slot j and waits for its turn, or Time Slot j, for the next frame. During other time slots, the MS is idle, i.e., has no access to the physical channel.

What would be the bit transmission rate of a channel? To determine this, consider the following analogy. Think of the bit positions passing by at the rate of $B_{max}/2$ b/s (1/2 factor is due to only one direction of transmission) as an infinite-length train with $B_{max}/2$ compartments passing by you every second. A time interval called a *time slot* is allocated to you periodically. During your time slot, N_{BP} compartments pass by you. Since N_{TS} time slots are grouped into one frame, you get your time slot every N_{TS}^{-th} time slot. In other words, you get your turn once every frame duration.

When you get your time slot, you can load your cargo on all of the compartments passing by you during that time slot. Once your time slot expires, you watch the compartments passing by but you cannot load your cargo on the train until your next time slot comes.

In this analogy, the speed of the train is the maximum allowable bit rate of the total air interface in one direction, which is $B_{max}/2$ b/s. Since the bit positions in a time slot pass by at this same speed, while you are loading your cargo during your time slot, you are transmitting at this maximum speed. However, once your time slot expires, you are idle until your next time slot. Therefore, during one frame duration, your are able to send only N_{BP} bits. Hence, even though the instantaneous bit rate during your time slot is same as the maximum air interface speed in one direction, the effective bit transmission rate in one direction of an individual channel, B_j, is:

$$\text{Reverse (or forward) link bit rate of individual channel} = B_j$$

$$= N_{BP} \text{ bits/frame duration}$$

$$= (N_{BP}) / \{(N_{TS})(N_{BP})(2/B_{max})\}$$

$$= (B_{max}/2) / (N_{TS}) \quad \text{b/s.}$$

The above round-about derivation is deliberate in order to show the relationship between various quantities. In fact, the above relation could have been derived by the following simple observation. The maximum bit rate for the total air interface in one direction is $B_{max}/2$. There are N_{TS} time slots per frame yielding N_{TS} individual channels. Since the air interface is shared by N_F individual channels, each channel can transmit at the most at the rate of $(1/ N_F)^{-th}$ of the maximum bit rate of $B_{max}/2$, which is $(B_{max}/2) / (N_{TS})$ b/s.

The TDD/TDMA

In the FDD/TDMA method discussed above, the two transmission directions are first created by dividing the total spectrum W into two halves, and the bit positions (BP's) in a time slot are used for one direction only. In the TDD/TDMA method, the maximum bit rate for the total air interface for both directions, B_{max}, is divided among the total number of duplex physical channels, and the reverse and forward directions for a given channel are created by dividing the bit rate allocated for each channel between the two directions.

Let:

Number of BP's in a time slot in the FDD/TDMA (and TDD/TDMA)

$$= N_{FDD-BP} \text{ (and } N_{TDD-BP})$$

Number of time slots in a frame in the FDD/TDMA (and TDD/TDMA)

$$= N_{FDD-TS} \text{ (and } N_{TDD-TS})$$

If the efficiency of using the finite resource – the total bandwidth W – is assumed to be the same in the FDD and the TDD, both the FDD/TDMA and the TDD/TDMA must produce the same number of physical channels, i.e., time slots, with the same maximum bit rate in each transmission direction. Under this ideal comparison where the bandwidth utilization efficiency is kept the same, the following relationships hold true:

Number of time slots per frame: $N_{TDD-TS} = N_{FDD-TS}$

Number of BP's per time slot: $N_{TDD-BP} = (2)(N_{FDD-BP})$

Recall that the FDD frame contains BP's used in one direction only, whereas the TDD frame contains BP's used in both directions: one half of N_{TDD-BP} BP's are used for the reverse direction and the other half, for the forward direction. Typically, within a given time slot, the BP's may be used alternately between the two directions or may be divided into two consecutive halves of BP's. These two methods are shown in Figure 2.7.

Since the time slot duration is the same in both the TDD and the FDD and the number of BP's is twice in the TDD as that in the FDD, the inter-BP time in the TDD is one half of that of the FDD:

Inter-BP time $\quad t_{TDD-BP} = (1/2)(t_{FDD-BP}) = 1/B_{max}$

Since only one half of the BP's in a TDD/TDMA time slot are used for one direction, the average inter-BP time in each direction is twice that shown above:

Inter-BP time in each direction

$$t_{TDD-R-BP} \text{ (or } t_{TDD-F-BP}) = t_{FDD-R-BP} \text{ (or } t_{FDD-F-BP})$$

$$= 2/B_{max} \text{ second}$$

Time slot duration

$$t_{\text{TDD-TS}} \quad = \quad t_{\text{FDD-TS}}$$

Frame duration

$$t_{\text{TDD-F}} \quad = \quad t_{\text{FDD-F}}$$

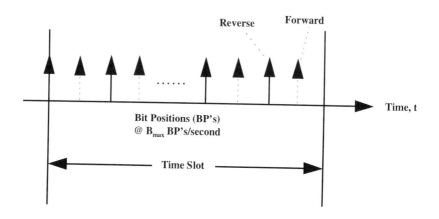

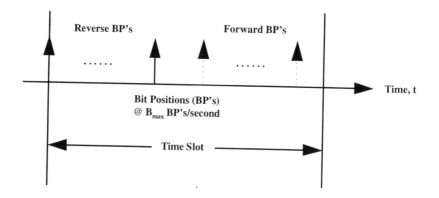

Figure 2.7 The reverse and forward bit positions in the TDD

The bit transmission rate of a TDD/TDMA channel in each direction is:

Reverse (or forward) link bit rate of individual channel, B_j

$$= N_{BP} \text{ bits/frame duration}$$

$$= (N_{TDD-BP}/2) / \{(N_{TDD-TS})(N_{TDD-BP})(1/B_{max})\}$$

$$= (B_{max}/2) / (N_{TDD-TS}) \quad b/s.$$

In the above, $N_{TDD-BP}/2$ is used because only one half of N_{TDD-BP} are the BP's used in one direction. This channel bit transmission rate per direction is same as that in the FDD/TDMA discussed earlier noting that $N_{TDD-TS} = N_{FDD-TS}$.

The Code Division Multiple Access (CDMA)

Recall the equivalence between the total bandwidth W Hz and the maximum bit rate B_{max} b/s of the air interface. In the Code Division Multiple Access (CDMA) system, all the physical channels on the air interface are allowed to transmit at this maximum bit rate, B_{max}, at the air interface. At the user interface, however, the individual channels have the bit rate lower than that over the air. Let the individual channel bit rate be, B_j. Each channel boosts this bit rate to the same maximum possible rate over the air, B_{max}. The maximum bit rate over the air in the CDMA system is referred to as the *chip rate*.

If all the physical channels transmit over the air at the same maximum bit rate B_{max}, the frequency spectrum of each channel would occupy the total bandwidth W. How is it then possible to separate the individual channels? This separation of channels is possible by means assigning unique codes to individual channels. As the individual channel boosts its bit rate to B_{max}, it uses the unique assigned code bit string to append to each bit of the original bit string. At the receiving end, each channel has a decoder, which recognizes its unique code pattern. To the decoder of a given channel, other signals appear as a background noise.

3

THE STANDARD BASED ON THE NORTH AMERICAN HIGH-TIER TDMA SYSTEM

3.1 OVERVIEW

This wireless access system for the 1.9 GHz PCS is derived from the existing North American 900-MHz TDMA cellular standard referred to as the IS-136 standard[6]. The following are some of the main characteristics of this system:

■ *Logical channel structure.* One major enhancement provided by this digital system as compared to the analog cellular system, the Advanced Mobile Phone Service (AMPS), is the addition of the Digital Control Channel (DCCH), which is a collection of logical channels conveyed on radio bearer channels. The DCCH is divided into the reverse DCCH (RDCCH) and the forward DCCH (FDCCH). The RDCCH in turn consists of a Random Access Channel (RACH). The FDCCH consists of: Short Message Service (SMS) Point-to-Point, Paging and Access Response Channel (SPACH), Broadcast Control Channel (BCCH), Shared Channel Feedback (SCF) and reserved time slots. The SPACH consists of the following subchannels: Paging Channel (PCH), Access Response Channel (ARCH), and SMS Channel (SMSCH). The BCCH consists of the following subchannels: Fast Broadcast Control Channel (F-BCCH), Extended Broadcast Control Channel (E-BCCH), and SMS Broadcast Control Channel (S-BCCH).

■ *Multiple access and time slot structure.* The multiple access scheme is TDMA. A TDMA frame is 40 milliseconds long and consists of six equally sized time slots, each 162 symbols (324 bits) in length with six time slots per frame. The channel bandwidth is 30 kHz.

■ *Authentication.* The authentication process involves updating the Shared Secret Data (SSD). The SSD is a 128-bit pattern stored in the mobile station and is readily available to the base station. The SSD is partitioned into two portions: SSD-A and SSD-B. SSD-A is used to support the authentication procedure and SSD-B, voice privacy and message confidentiality. A successful outcome of the authentication process occurs only when it can be demonstrated that the mobile station and the base station possess identical sets of Shared Secret Data (SSD). This is done through a series of challenge-response steps between the mobile station and the base station using the CAVE algorithm. This type of authentication procedure is sometimes referred to as a private key method.

■ *Handoff.* This standard uses the Mobile Assisted Handoff (MAHO). The MAHO function requires a mobile station to furnish RF-channel signal quality information to its serving base station. The mobile station starts channel quality measurements with the reception of the "start measurements" order from the base station. This message identifies those forward RF channels which the base station requires the mobile station to measure. The mobile station transmits the measurements over either the SACCH or FACCH. Based on these measurements, the network then selects a new channel.

This system consists of two groups of logical channels referred to as the Digital Control Channel (DCCH) and the Digital Traffic Channel (DTC), respectively. The DCCH is a group of logical channels used for the signaling between the base and mobile stations during the registration, paging, call delivery, call origination and other call processing procedures. The DCCH is also used for transmission of short user data messages between the base and mobile stations, and allows the capability of providing the Short Message Services. The DTC is a group of logical channels used for user calls. In addition, the DTC contains control channels for supervisory signaling between the base and mobile stations. Section 3.2 discusses the DCCH, and Section 3.3, the DTC.

3.2 THE DIGITAL CONTROL CHANNEL

3.2.1 The Protocol Reference Model

Figure 3.1 illustrates the three-layer protocol reference model used in this system. The base station (BS) and the mobile station (MS) communicate with each other at three layers of peer-to-peer protocol: Layer 3, Layer 2, and Layer 1, which is also referred to as the Physical Layer in this system. The protocols are defined in terms of the sets of message and the procedures. The Layer 3 protocols are defined for the messages and procedures necessary for the signaling beyond the air interface such as those for the registration, call origination, call termination, and hand-off procedures. Layer 2 protocols are primarily concerned about the reliable transmission of the information between the two entities, i.e., the BS and the MS. Layer 1 puts the information into the physical channel.

Within each entity such as the BS and the MS, each layer provides the services to the layer above it and relies on the services provided by the layer below it. Layer 3 receives the services from Layer 2. Layer 3 is the highest layer specified in this system. It provides the services to the user interface and receives requests from the user. Layer 2 provides the services to Layer 3 and relies on the services provided by Layer 1. Similarly, Layer 1 provides the services to Layer 2 and interfaces with the physical channel.

The logical point of interface or address at which one layer communicates with the layer above it is referred to as the Service Access Point (SAP). For example, for a given Layer 2 service, the Layer 2 SAP is the point at which Layer 2 receives a request for the service from, and delivers the service to, Layer 3. Similarly, a Layer 1 SAP is the point at which Layer 1 receives a request for its service from, and delivers the service to, Layer 2.

Within an entity, the communications between the layers are accomplished by using the internal messages, referred to as the primitives, as opposed to the protocol messages, which are used by the peer-to-peer protocols between two different entities. Two types of primitives are used in this system: the request primitives and the indication primitives. The request primitives are sent by one layer, say Layer N, to the lower layer, Layer (N-1); and the indication primitives, sent by Layer N to the high layer, Layer (N+1). For example, Layer 3 sends request primitives to Layer 2. Layer 2 sends the request primitives to Layer 1 and the indication primitives to Layer

3. Layer 1 sends indication primitives to Layer 2. Unlike Layers 2 and 3, Layer 1 does not use the request primitives: this layer puts the information into the physical channel formats without any explicit use of the request primitives. This three-lay protocol reference model will be referred to as we proceed with the discussion of the logical channels.

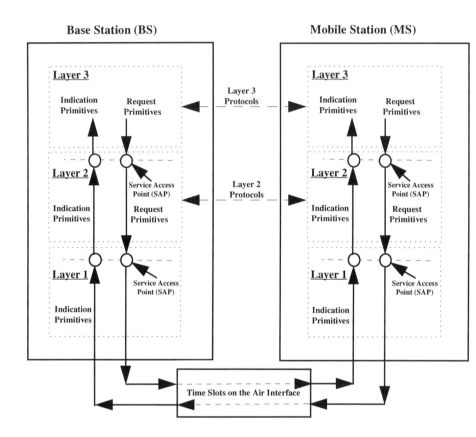

Figure 3.1 The Three-Layer Protocol reference Model

3.2.2 The DCCH Logical Channels

The Digital Control Channel (DCCH) is divided into the Reverse Digital Control Channel (RDCCH) and the Forward Digital Control Channel (FDCCH). As the descriptors, "reverse" and "forward," suggest, the RDCCH and the FDCCH are unidirectional, and, therefore, carry messages in one direction only. If a message transmitted in the RDCCH requires an acknowledgment, the acknowledgment is sent in the FDCCH, and *vice versa*. The RDCCH consists of only one type of logical channel, the Random Access Channel (RACH). The RACH is used by the MS to request a contention or reservation based access to the system. The messages transmitted on the RACH require acknowledgments from the base station and the acknowledgments from the base station are transmitted to the MS in one particular type of logical channel of the FDCCH.

The FDCCH consists of four types of logical channels: the Short Message Service (SMS) Point-to-Point, Paging and Access Response Channel (SPACH), the Broadcast Control Channel (BCCH), the Shared Channel Feedback (SCF) channel, and the reserved channel.

The SPACH is used to broadcast information to specific mobile stations and is further subdivided into the following three logical channels depending on the type of the information: the Paging Channel (PCH), the Access Response Channel (ARCH), and the SMS Channel (SMSCH). The Paging Channel (PCH) is used to broadcast pages and orders. The messages transmitted on the PCH do not require acknowledgments. Since the PCH operates in the unacknowledged mode, it does not require the Automatic Retransmission Request (ARQ) protocol.

The Access Response Channel (ARCH) is used by the MS when the MS is expecting a response to certain access attempts on the RACH. The MS autonomously moves to the ARCH after successfully accessing an RACH. The ARCH operates either in the unacknowledged mode or the acknowledged mode. Since the ARCH may operate in the acknowledged mode, it supports the ARQ protocol. The SMS Channel (SMSCH) is used to deliver short messages to a specific mobile station as in the SMS services. The SMSCH operates either in the unacknowledged mode or the acknowledged mode. Since the SMSCH may operate in the acknowledged mode, it supports the ARQ protocol.

The BCCH is used to broadcast generic, system-related information from the base station to the MS's and operates in the unacknowledged mode. The BCCH is further divided into three subchannels: the Fast Broadcast Channel (F-BCCH), the Extended Broadcast Channel (E-BCCH) and the SMS Broadcast Channel (S-BCCH). The F-

BCCH is used to broadcast DCCH structure parameters and other parameters essential for accessing the system. The E-BCCH is used to broadcast less time-critical information. The S-BCCH is used for the broadcast SMS service.

The Shared Channel Feedback (SCF) channel is used in support of the RACH operation. The RACH, which is reverse logical channel, operates in the acknowledged mode, and the acknowledgments, which must travel in the forward direction, are transmitted on the SCF, which is a forward DCCH. Finally, the Reserved Logical Channel is reserved for future use.

The following figure summarizes the logical channels of the Digital Control Channel:

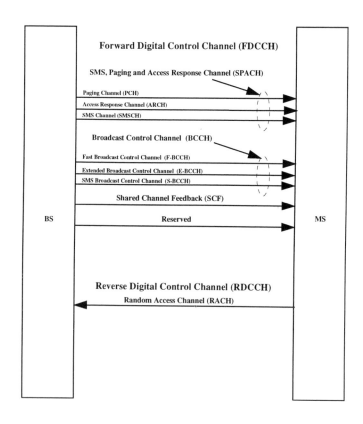

3.2.3 Layer 3 Message Segmentation and Re-assembly

Figure 3.2 illustrates the segmentation of a Layer 3 message. Each Layer 3 message is segmented into N segments; each segment is framed into a Layer 2 (L2) frame; and each L2 frame is appended with tail (T) bits.

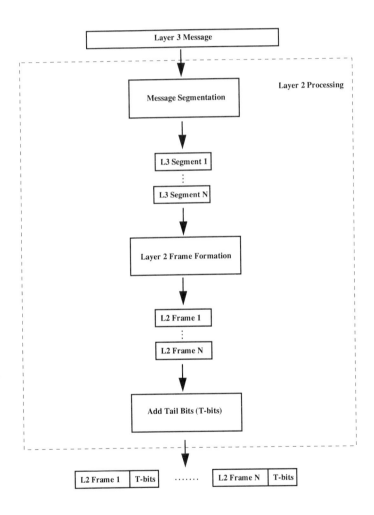

Figure 3.2   The segmentation of a Layer 3 message for the mapping to Layer 2

Figure 3.3 illustrates the mapping of the L2 frames to Layer 1 for transmission in time slots. As shown in Figure 3.2, the Layer 3 message is first segmented into Layer 2 (L2) frames. The L2 frames are appended with T-bits. Each L2 frame plus T-bits is then subjected to the channel coding and interleaving. The output of the channel coding and interleaving process is put into the time slot for transmission over the air interface.

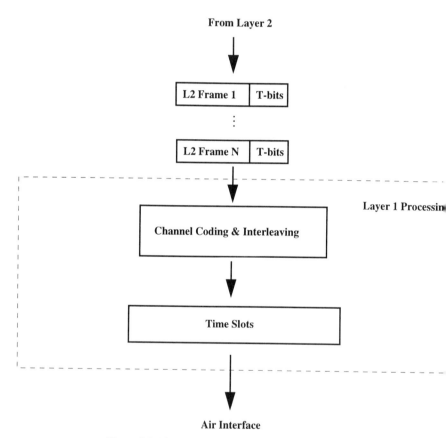

Figure 3.3 The mapping of the Layer 2 to Layer 1

Figures 3.4 and 3.5 illustrate decoding and de-interleaving of L2 frames and re-assembly of the L2 messages into the Layer 3 message as transmitted from the other end. These two figures show the reverse process of the segmentation, channel coding and interleaving shown in Figures 3.2 and 3.3. The L2 frames appended with the T-bits are first extracted from the time slot through the Layer 1 processing. These L2 frames plus T-bits are passed to the Layer 2 processing. In the Layer 2 processing, the T-bits are removed from the L2 frames. From the L2 frames, the Layer 3 (L3) message segments are extracted, which are then re-assembled to the Layer 3 messages sent from the transmitting end.

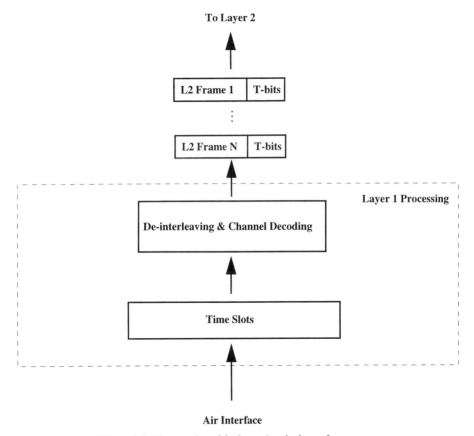

Figure 3.4 The mapping of the Layer 1 to the Layer 2

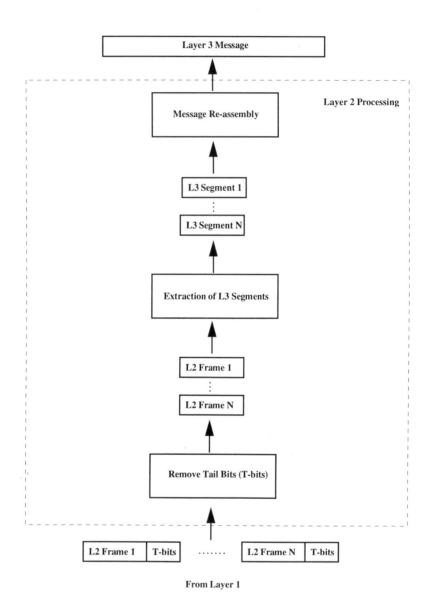

Figure 3.5   The re-assembly of a Layer 3 message

3.2.4 The Physical Channels

The distinction between the physical channel and the logical channel is not always clear-cut. A pair of wires which connects your telephone set to the central office may be considered a physical channel. The radio channel frequency that you listen to may be another example of the physical channel. Loosely speaking, the more the structure of the channel is tied to the hardware design of the system, the closer the channel is to being a physical channel. A frequency, a time slot, etc. may be more of an integral part of the hardware design and may not easily be changed without changing the hardware. The logical channel may be defined as the logical application of the physical channel. For example, several frequency channels may be created as physical channels, and some of them may be used for paging and some of them, for traffic. The physical channels used for the paging may then be called the paging channels, and those used for traffic, the traffic channels, and both are logical channels.

In most wireless systems, physical channels are created by a combination of more than one techniques. The current system is not an exception. In this system, physical channels are created by a combination of the Frequency Division Duplexing (FDD), the Frequency Division Multiple Access (FDMA) and the Time Division Multiple Access (TDMA) techniques as described below.

The Creation of the Reverse and Forward Links by the Frequency Division Duplexing (FDD)

First, two separate frequency bands are assigned for the forward and reverse directions of transmission using the FDD technique. The FCC allocated a total of 120 MHz for the licensed PCS operation: 60 MHz of spectrum from 1850 to 1910 MHz, the low band of the A-, B-, C-, D-, E-, and F-blocks, and 60 MHz of spectrum from 1930 to 1990 MHz, the high band of the A-, B-, C-, D-, E-, and F-blocks. In this system, the low 60-MHz band, i.e., the 1810-1910 MHz band, is assigned for the reverse link and the high 60-MHz band, i.e., the 1930-1990 MHz band, for the forward link. Figure 3.6 illustrates the creation of the reverse and forward links by the FDD for this system.

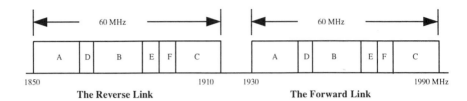

Figure 3.6   The frequency bands for the reverse and forward links created by the FDD technique

The Creation of the Frequency Channels by the Frequency Division Multiple Access (FDMA)

The 60-MHz-wide frequency band for each direction is then divided into smaller bands of frequency. These subbands of frequency are referred to as frequency channels. The 60-MHz band from 1810 MHz to 1910 MHz and that from 1930 to 1990 MHz are divided into 30-kHz-wide frequency channels with the center frequencies, f_N, given by the following equations:

Reverse link

$$f_N = 1850.010 + 0.030 \text{ MHz}, \qquad N = 1, 2, \ldots, 1999$$

Forward link

$$f_N = 1929.990 + 0.030 \text{ MHz}, \qquad N = 1, 2, \ldots, 1999$$

where N denotes the channel number. Channel 1 and Channel 1999 are not used.

A pair of these 30-kHz channels in the reverse and forward links constitutes one full duplex frequency channel.

The Creation of
the Time Slots by
the Time Division Multiple Access (TDMA)

Each 30-kHz frequency channel in each direction is divided into 48,600 bit positions per second yielding the total transmission capacity of the air interface of 48.6 kbps in each direction. These bit positions are grouped into frames. The duration of one frame is 40 ms, and thus each frame contains 1944 bit positions. These 1944 bit positions per frame are divided into six time slots of equal duration. Therefore, the duration of each time slot is 6.66.. ms, or 40/6 ms to be exact, and contains 324 bit positions. A pair of time slots in the reverse and forward links constitute one full duplex physical channel.

The six time slots are grouped into two blocks: Slots 1, 2 and 3 constituting one block, and Slots 4, 5 and 6, the other. As will be discussed further later, a full rate channel uses two time slots per frame, one from each block, i.e., Slots 1 and 4, or Slots 2 and 5, or Slots 3 and 6. A half-rate channel uses one slot per frame. Figure 3.7 shows the TDMA frame structure.

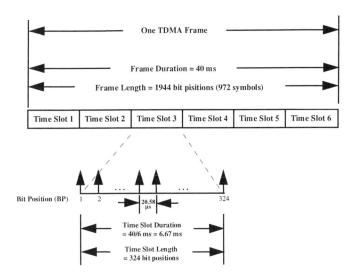

Figure 3.7 The TDMA frame structure

The physical layer, or Layer 1, provides services to Layer 2 at logical points of communication between the two layers referred to as the Service Access Points (SAP's). The physical layer receives the Layer 2 frames within the same entity, e.g., an MS, at appropriate SAP's depending on the logical channel used by the Layer 2 frames, performs the channel coding and interleaving operations on these frames and puts the results of these operations into the time slots for transmission to another entity, e.g., the BS. Conversely, the physical layer in one entity, e.g., an MS, receives Layer 2 frames from another entity, e.g., the BS, performs the reverse operations of the channel coding and interleaving and delivers the results to Layer 2 within the same entity, i.e., the MS, at the appropriate SAP's.

The Time Slot Formats

In a given time slot, the bit positions are assigned to carry different types of information according to the time slot formats. The time slot formats are different between the forward link, i.e., those carrying the FDCCH frames from the BS to the MS, and the reverse link, i.e., those carrying the RDCCH frames from the MS to the BS. For the FDCCH, only one format is used; however, for the RDCCH, two different time slot formats are used: normal and abbreviated formats.

Figure 3.8 shows the time slot format of the Reverse DCCH (RDCCH). The figure shows the names of the fields, the field lengths in bits and the bit position (BP) numbers of the bits in each field. The figure also shows the direction of transmission, i.e., from the MS to the BS. As discussed earlier, each time slot contains 324 BP's. The bit on BP number 1, i.e., the first bit of the G-field, is transmitted first and that on BP number 324, i.e., the last bit of the second data field in the normal format, or the last bit of the AG-field in the abbreviated format, is transmitted last. Figure 3.9 shows the time slot format of the DCCH on the forward link, i.e., the FDCCH.

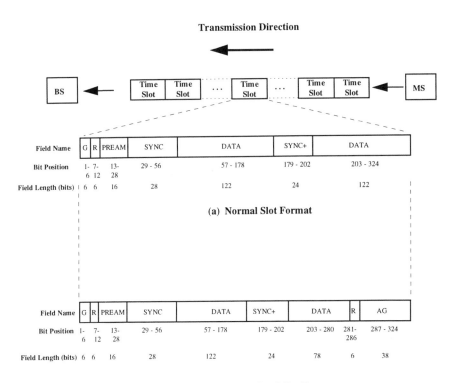

Transmission Direction

(a) Normal Slot Format

(b) Abbreviated Slot Format

Legends

G	- Guard Time	DATA	-	Coded and interleaved information bits
R	- Ramp Time	SYNC+	-	Additional Synchronization
PREAM	- Preamble	AG	-	Additional Guard Tinme
SYNC	- Synchronization			

Figure 3.8 The TDMA time slot structures on the Reverse Digital Control Channel (RDCCH)

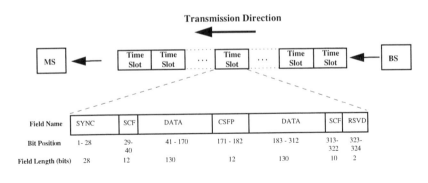

Legends

SYNC - Synchronization   CSFP - Coded Super Frame Phase
SCF   - Shared Channel Feedback   RSVD - Reserved Field (Bit Settings = "11")
DATA - Coded & interleaved information bits

Figure 3.9   The TDMA time slot structure on the Forward Digital Control Channel (FDCCH)

Referring to Figure 3.9, one of the logical channels defined for the Forward Digital Control Channel (FDCCH) is the Shared Channel Feedback (SCF). Unlike other logical channels, the SCF exists as a fixed field of the time slot structure of the FDCCH. The SCF field occupies 22 bit positions (BP's) numbered from 29 through 40 and from 313 through 322. These 22 BP's are divided into three subfields as follows:

■   Busy/Reserved/Idle (BRI)

■   Received/Not Received (R/N)

■   Coded Partial Echo (CPE)

These three subfields of the SCF field are shown in Figure 3.10.

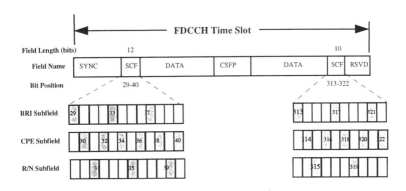

Figure 3.10   The subfields of the SCF field of the FDCCH time slot

As illustrated in Figure 3.3, the Layer 2 frames are appended with T-bits and the resulting bits are subjected to the channel coding and interleaving. The output of the channel coding and interleaving is then put into the time slots. These bits are put into the DATA-fields of the time slots whose formats are shown in Figures 3.8 and 3.9. In order to understand this mapping between Layer 2 and Layer 1, i.e., the time slots, we need to understand the Layer 3 message segmentation process along with the Layer 2 frame structure. Furthermore, we need to understand the channel coding and interleaving to the extent that we can understand the mapping. The channel coding and interleaving will be discussed further later. First, we will discuss the Layer 2 frame structure, the channel coding and interleaving, and the layer 2 to Layer 1 mapping.

Figure 3.11 shows the Layer 2 frame structure. As will be seen shortly, the total length of the Layer 2 frame should be such that, after appending the Layer 2 frame with the T-field of five bits and subjecting the resulting bits to the channel coding and interleaving process, the output bits of the process should fit into the DATA-fields of the time slot. Therefore, the lengths of the Layer 2 frame are fixed for the logical channels. They are 125 bits for the SPACH and the BCCH, i.e., the forward channels, and 117 and 95 bits for the normal length and abbreviated length RACH frames, respectively. The reason for these specific lengths will become clear when we discuss the channel coding.

Since the CRC field has a fixed length of 16 bits, the combined length of the Layer 2 Header field and the Layer 3 Message Segment field are 108 bits for the SPACH and the BCCH, and 101, and 79 bits for the normal length and abbreviated length RACH frames, respectively. The length of the Layer 2 header is variable. Therefore, for a given Layer 2 Header length, the length of the Layer 3 Message Segment is determined by subtracting the Layer 2 Header length from these combined lengths.

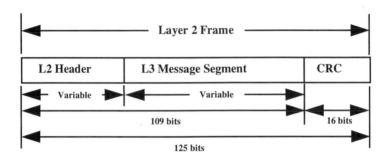

(a) SPACH and BCCH

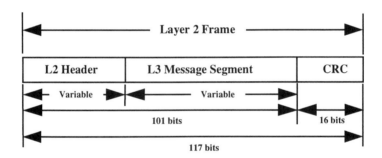

(b) RACH - Normal Length

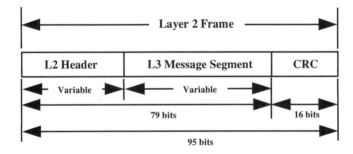

(c) RACH - Abbreviated Length

Figure 3.11   The frame structure of Layer 2 frames on the SPACH, BCCH and RACH

The input to the channel coding is the Layer 2 frame plus the five bits of the T-field. For the SPACH and the BCCH, the total number of bits of one Layer 2 frame plus the T-field is 130. Similarly, the total number of bits of one Layer 2 frame plus the T-field is 122 for the normal length RACH frame, and 100 for the abbreviated length RACH frame. Now, both the RDCCH and the FDCCH use the convolutional encoding with code rate r = 1/2. This means that the bit rate doubles after the encoding. The interleaving does not change the bit rate. Hence, after the channel coding and interleaving, the one Layer 2 frame plus the T-field results in 260, 244 and 200 bits for the SPACH/BCCH, normal RACH, and abbreviated RACH, respectively. These bits fit into the DATA-fields of the corresponding time slots, respectively.

Figures 3.12 through 3.14 illustrate these mappings for the SPACH/BCCH, normal RACH and abbreviated RACH.

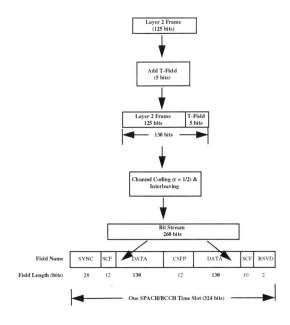

Figure 3.12 The mapping of Layer 2 frames to the DATA fields of the time slot on the SPACH and the BCCH

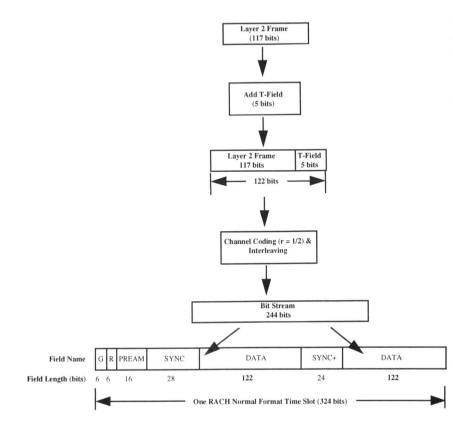

Figure 3.13 The mapping of Layer 2 frames to the DATA fields of the time slot on the normal format RACH

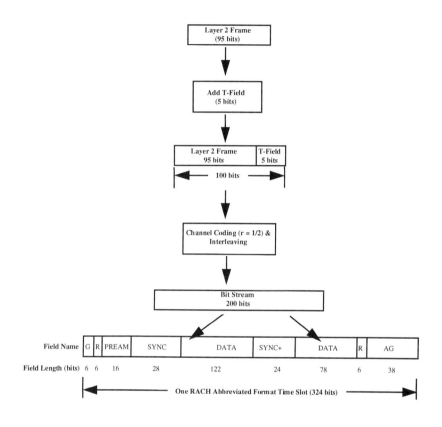

Figure 3.14   The mapping of Layer 2 frames to the DATA fields of the time slot on the abbreviated format RACH

The Superframe Format

Recall that the FDCCH consists of the following four types of logical channels: SPACH, BCCH, Reserved, and SCF. The superframe is used to divide the FDCCH into these logical channels. The use of the superframe for creating these logical channels is discussed next.

The full-rate FDCCH has a cycle of two time slots repeating in every TDMA frame of six time slots, with the frame duration of 40 ms. The two slots are the first time slots of the two TDMA blocks of the frame, i.e., Slot 1 and Slot 4. The total number of bit positions (BP's) contained in one TDMA frame is 1944, and that in one time slot, 324. Since the full-rate FDCCH transmits the bits contained in two time slots in every 40 ms, the bit transmission rate of the full-rate FDCCH is:

$$(324\ bitsx2)/(40\ ms) = 16.2\ kb/s$$

Similarly, the half-rate FDCCH has a cycle of one time slot, Slot 1, repeating in every TDMA frame of six time slots. Therefore, the bit transmission rate of the half-rate FDCCH is:

$$(324\ bits)/(40\ ms) = 8.1\ kb/s$$

One superframe is constructed by collecting the FDCCH time slots in sequence from 16 TDMA frames, or, since each TDMA frame contains two TDMA blocks of three time slots each, equivalently 32 TDMA blocks. The bit transmission rate of the superframe is same as that of the FDCCH: the full-rate superframe operates at 16.2 kb/s and the half-rate superframe, at 8.1 kb/s.

The superframe is divided into 32 slots with the superframe phase (SFP) number ranging from 0 to 31. These 32 SFP slots correspond to the 32 TDMA blocks from which the FDCCH time slots are extracted. The superframe duration is equal to the duration of 16 TDMA frames and the duration of one SFP slot is equal to one thirtieth of the superframe duration, or same as that of one TDMA block:

$$Duration\ of\ one\ superframe = 40\ ms\ x\ 16 = 640\ ms$$
$$Duration\ of\ one\ SFP\ slot = 640\ ms/32 = 20\ ms.$$

Figure 3.15 illustrates the mapping of the full-rate FDCCH time slots into the superframe. The FDCCH time slots are put into the SFP slots, one time slot per SFP slot. Therefore, the number of bit positions (BP's) in one SFP slot is same as that in one time slot, i.e., 324 BP's, and the time between two consecutive BP's in a superframe. The two FDCCH time slots – one per TDMA block – in each TDMA frame are mapped into two consecutive SFP slots. The bits contained in one FDCCH time slot fills one SFP slot. Since the number of bits contained in one FDCCH time slot is same as that contained in one SFP slot, whereas the duration of one SFP slot is three times that of one time slot, the time distance between two bits in the SFP slot is three times that in the time slot, i.e., 20 ms/324 = 61.73 µs.

The 324 bits contained in one time slot are repeated in every three time slots in the TDMA frame and the same number of bits, 324 bits, are repeated in every SFP slot, whose duration is same as that of three time slots in the TDMA frame. The same number of bits are repeated in the same amount of time in both the TDMA frame and in the superframe. Hence, the bit transmission rate of the full-rate FDCCH, 16.2 kb/s, is achieved by the superframe constructed in this manner.

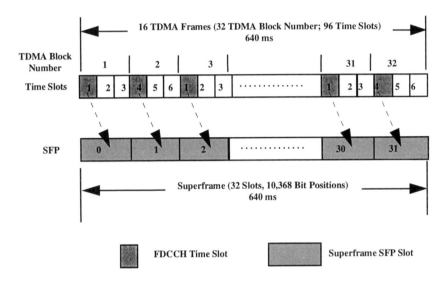

Figure 3.15 The mapping of full-rate FDCCH time slots unto the superframe slots

Figure 3.16 illustrates the mapping of the half-rate FDCCH time slots into the superframe. In this case, each TDMA frame contains only one FDCCH time slot, i.e., only the first TDMA block contains the FDCCH time slot. The FDCCH time slots are mapped into even-numbered SFP slots only. Since the SFP slots containing the FDCCH time slots repeat themselves every other slot, the half-rate FDCCH bit transmission rate, 8.1 kb/s, is achieved.

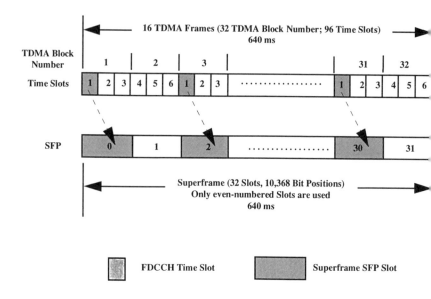

Figure 3.16 The mapping of half-rate FDCCH time slots unto the superframe slots

Having constructed the structure of the superframe as discussed above, the logical channels of the FDCCH are then created by assigning them to the SFP slots. Figures 3.17 and 3.18 illustrate the logical channel assignments in the superframe for the full- and half-rate FDCCH, respectively. The BCCH, Reserved and SPACH channels are assigned to the SFP slots consecutively in that order. The BCCH slots are then further divided into its sub-channels – F-BCCH, E-BCCH and S-BCCH – also consecutively. An SFP slot in the superframe repeats itself every 640 ms, and the number of SFP slots assigned for a given logical channel determines the bit transmission rate of that channel, i.e., more SFP slots, faster the logical channel speed. The minimum and maximum number of SFP slots that can be assigned for the logical channels are specified in the standard. These minimum and maximum number of slots are also shown in the figures and Table 3.1 shows the corresponding minimum and maximum bit transmission rates of these logical channels.

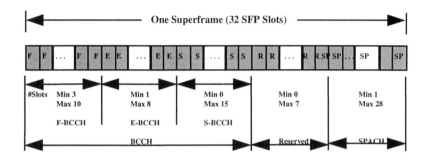

Figure 3.17 The full-rate logical channel sequence in the superframe

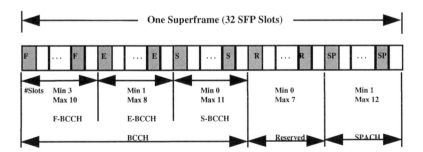

Figure 3.18 The half-rate logical channel sequence in the superframe

FDCCH Logical Channels	Min Num of Slots	Max Num of Slots	Min. Trans Rate (kb/s)	Max Trans Rate (kb/s)
F-BCCH	3	10	1.52	5.06
E-BCCH	1	8	0.51	4.05
S-BCCH	0	15	0	7.59
Reserved	0	7	0	3.54
SPACH	1	28	0.51	14.18

(a) Full-Rate FDCCH

FDCCH Logical Channels	Min Num of Slots	Max Num of Slots	Min. Trans Rate (kb/s)	Max Trans Rate (kb/s)
F-BCCH	3	10	1.52	5.06
E-BCCH	1	8	0.51	4.05
S-BCCH	0	11	0	5.57
Reserved	0	7	0	3.54
SPACH	1	12	0.51	6.08

(a) Half-Rate FDCCH

Table 3.1 The total transmission rates and the effective information transmission rates of the logical channels

What about the SCF channel? Recall that the SCF channel is a fixed field of every FDCCH time slot. The SCF channel takes up a total of 22 bits in every time slot of 324 bits. Therefore, each SFP slot containing a FDCCH time slot also contains 22 bits of SCF channel. This is illustrated in Figure 3.19. Since the duration of one SFP slot is 20 ms, the bit transmission rate of the SCF channel is 22 bits/20 ms = 1.1 kb/s.

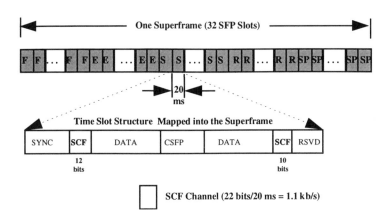

Figure 3.19   The SCF channel in the superframe

Example

We now consider an example of the mapping of the FDCCH time slots to the superframe for specific assignments of the logical channels to the SFP slots. Suppose that the following are the numbers of SFP slots assigned for the logical channels for the full- and half-rate FDCCH's. The corresponding bit transmission rates are also shown. Note that the sum of the numbers of SFP slots and the sum of the bit transmission rates of the individual logical channels must agree with the total number of SFP slots and the total bit transmission rate of the FDCCH, i.e., 32 and 12 slots and 16.2 and 8.1 kb/s for the full- and half-rate FDCCH, respectively. Figures 3.20 and 3.21 show the complete mapping from the FDCCH time slots of the TDMA frame to the superframe for the full- and half-rate FDCCH's, respectively.

FDCCH Logical Channels	*Full-Rate FDCCH*		*Half-Rate FDCCH*	
	Number of Slots	*Max Trans Rate (kb/s)*	*Number of Slots*	*Max Trans Rate (kb/s)*
F-BCCH	4	2.02500	4	2.02500
E-BCCH	1	0.50625	1	0.50625
S-BCCH	2	1.01250	2	1.01250
Reserved	3	1.51875	3	1.51875
SPACH	22	11.13750	6	3.03750
Total	32	16.2	16	8.1

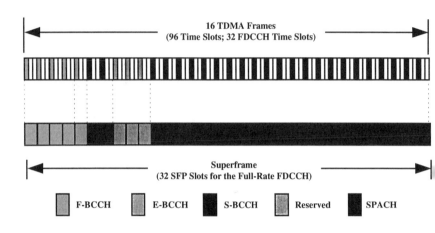

Figure 3.20 The mapping of the full-rate FDCCH to the superframe for the example

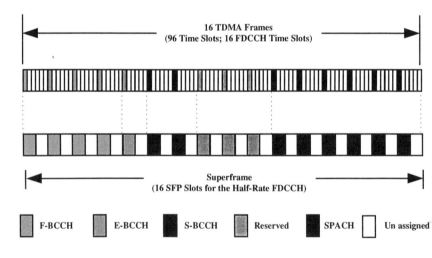

Figure 3.21 The mapping of the half-rate FDCCH to the superframe for the example

The Hyperframe Format

The hyperframe is used to control the periodicity of the content of the logical channel frames, and consists of two superframes: one primary and one secondary.

3.2.5 Layer 2 Protocols

The RACH Protocol

The Layer 2 frame structure

The RACH is the only logical channel on the RDCCH. The Layer 2 frame for the RACH carries Layer 3 messages used on that logical channel. The Layer 2 frame is carried in the DATA-field of the RDCCH time slot. As discussed earlier, a Layer 2 frame is first appended with five bits of the tail (T)-field and are encoded to produce a bit stream twice the combined length of the Layer 2 frame and the T-field. This bit stream after the encoding must fit in the DATA-field of the RDCCH time slot.

Since the encoding process doubles the number of bits, the length of the Layer 2 (L2) frame is calculated as follows:

(L2 frame length + 5 bits of T-field) x 2 = DATA-field length, or

L2 frame length = DATA-field length/2 -5

The length of the DATA-field of a normal-length RDCCH time slot is 244 bits, and that for the abbreviated-length RDCCH time slot, 200 bits. Substituting these numbers in the above:

$$L2\ frame\ length \quad = \quad 117\ bits \quad normal\ time\ slot\ format$$
$$= \quad 95\ bits \quad abbreviated\ time\ slot\ format$$

There are five different types of L2 frames used on the RACH: BEGIN frame, CONTINUE frame, END frame, BEGIN and END combined frame, and SPACH ARQ STATUS frame. The first four are used to carry Layer 3 (L3) messages. The last is used to support the ARQ protocol used on the SPACH on the FDCCH. Since all logical channels are unidirectional, in order to implement the ARQ protocol on the FDCCH, there must be a communications path in the reverse direction. The

RACH is serving this purpose for the SPACH ARQ protocol, and the last L2 frame listed above carries the ARQ status information for the ARQ protocol used on the SPACH.

The use of the first four types of L2 frames depends on the length of L3 messages and also on the number of L3 messages to be sent. If a single L3 message is short enough to fit into one L2 frame, the BEGIN and END combined frame is used; otherwise, the BEGIN-, CONTINUE- and END-frames are used in sequence. Figure 3.22 shows the general format of the RACH L2 BEGIN and BEGIN/END Combined frames for the normal and abbreviated time slot format, respectively. The following fields are used in the RACH L2 BEGIN and BEGIN/END Combined frames. The first six fields constitute the header of the frame. The lengths of the header field and the CRC field are the same in both the normal and abbreviated time slot formats. Only the Layer 3 data (L3DATA) field has different lengths between the two formats.

- *Burst Type (BT)*: The BT-field is present in all five types of RACH L2 frame and the field length is always three bits. The bit settings of this field determines which of the five types the current L2 frame is. The bit-settings are as follows:

BT-Field Bit-Settings	Frame Type
000	BEGIN
001	CONTINUE
010	END
011	BEGIN and END Combine
100	SPACH ARQ STATUS
101 . . . 111	Not used (i.e., reserved)

- *Identify Type (IDT)*: The IDT-field determines the type of Mobile Station (MS) identify. The field length is two bits. The bit-settings are as follows:

IDT-Field Bit-Settings	MS ID Type
00	20-bit TMSI
01	24-bit TMSI
10	34-bit MIN
11	50-bit IMSI

Because the MS identification requires different numbers of bits as seen above, the IDT-field bit-setting determines the length of the MSID field.

■ *Extension Header Indicator (EHI):* The EHI-field has the field length of one bit and indicates the presence (EHI = 1) or absence (EHI = 0) of the Extension Header. The Extension Header contains further header information in addition to the normal header comprised of the six fields, and is present when the Layer 3 information payload needs to be encrypted.

■ *Mobile Station Identity (MSID):* This field contains the identity of the MS. The field length varies with the bit-setting of the IDT-field. As shown in the figure, the MSID field length is 20, 24, 34 or 50 bits corresponding to the IDT-field bit-setting of 00, 01, 10 and 11, respectively.

■ *Number of Layer 3 Messages (NL3M):* The field length is three bits. The binary bits in this field are converted to a decimal number. The decimal number plus one indicates the number of Layer 3 messages contained in the current L2 frame. For example, all zeros in this field indicate one Layer 3 message, 001 indicates two Layer 3 messages, etc. Up to eight Layer 3 messages can be contained in one L2 frame.

■ *Layer 3 Length Indicator (L3LI):* The field length is eight bits. The decimal number corresponding to the binary bits in this field indicates the length of the Layer 3 message in octets. The maximum Layer 3 message length supported is 255 octets, which correspond to all 1's in the L3LI-field. If the Layer 3 messages are short so that more than one of them may fit into one L2 frame, additional L3LI's are used. The L3LI- field occurs in the same L2 frame contiguously as many times as the number of Layer 3 messages contained in the same L2 frame as indicated by the NL3M-field. Since the NL3M-field supports up to eight Layer 3 messages, up to eight instances of L3LI may occur in one L2 frame, i.e, L3LI1, L3LI2, ... , L3LI8. All of these L3LI fields are eight bits long.

■ *Layer 3 Data (L3DATA):* The L3DATA-field contains the Layer 3 message segments. This is the only field which has different lengths between the normal and abbreviated time slot format shown in Figures 3.22. Since the L2 frame for the abbreviated time slot is shorter than that for normal time slot by 22 bits, the L3DATA-field of the former is 22 bits shorter than that of the latter. When more then one L3LI-field occurs, the L3DATA field length is reduced by the lengths of the additional L3LI fields. For example, if the MSID-field length is 50 bits, i.e., the MSID type is the 50-bit IMSI, and two Layer 3 messages are contained in the L2 frame so that two L3LI indicators are included, the L3DADTA-field length is reduced by eight bits to 26 bits and four bits from 34 bits and 12 bits shown in Figures 5.11 and 5.12, respectively.

■ *Cyclic Redundancy Check (CRC):* 16 bits are used for the CRC. This field is present in all five types of RACH L2 frames.

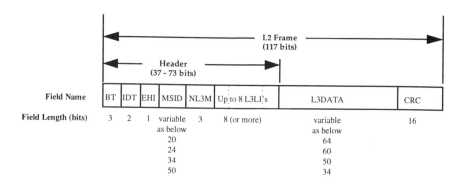

(a) For the Normal Time Slot

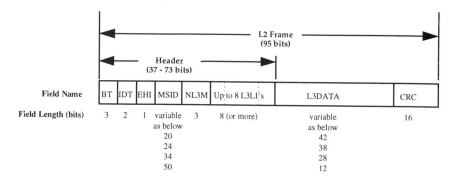

(b) For the Abbreviated Time Slot

Figure 3.22   General format of the RACH Layer 2 BEGIN and BEGIN/END Combined frames

Figure 3.23 shows the CONTINUE frame format. The Change Indicator (CI) field has the length one bit. It starts at 0 and toggles for every new transmitted frame and stays the same for every repeated frame.

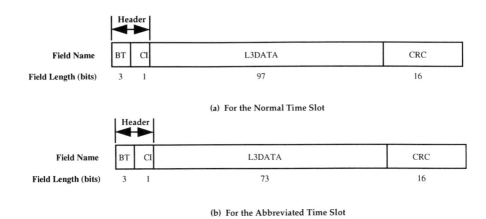

(a) For the Normal Time Slot

(b) For the Abbreviated Time Slot

Figure 3.23 General format of the RACH Layer 2 CONTINUE frame

Figure 3.24 shows the END frame format. The END frame format is same as the CONTINUE frame format except that the CI-field of the latter is replaced by the reserved (RSVD) field, which is set to 0.

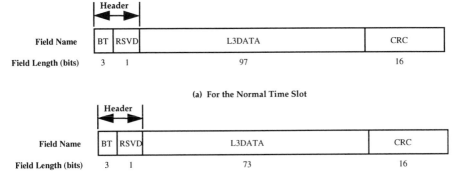

(a) For the Normal Time Slot

(b) For the Abbreviated Time Slot

Figure 3.24 General format of the RACH Layer 2 END frame

Examples

To illustrate the use of the different types of RACH L2 frames, several examples are given here. The actual bit-settings are given within the fields.

Case 1 – A single 32-bit (4-octet) L3 message is transmitted in one L2 frame for the normal time slot: NL3M = 000 (i.e., one L3 message); L3LI = 00000100 (i.e., 4 octets). Assume that the MSID type is MIN and no Extension Header is present: IDT = 10; EHI = 0. MIN is 34 bits long and the unique number for the MS. Since only one L2 frame is needed, the BEGIN/END Combined frame is used: BT = 011. The total length of the L3DATA-field for this type frame for the normal time slot is 50 bits. Thirty two bits, i.e., 4 octets, of these are L3 message data and the remaining 18 bits are all zeros as the filler. The resulting L2 frame is as shown by the following figure.

Field Name	BT	IDT	EHI	MSID	NL3M	L3LI	L3DATA	FILLER	CRC
Bit-Setting	011	10	0	XX..XX	000	00000100	X X	000.........000	XX......XX
Field Length (bits)	3	2	1	34	3	8	32 (4 octets)	18	16

BEGIN/END Combined Frame for the Normal Time Slot

Case 2 – A single 10-octet L3 message is transmitted in multiple L2 frames for the normal time slot: NL3M = 000 (i.e., one L3 message); L3LI = 00001010 (i.e., 10 octets). Assume that the MSID type is 20-bit TMSI and no Extension Header is present: IDT = 00; EHI = 0. Since the L3 message is 10 octets long, at least two L2 frames are needed. One BEGIN frame takes 64 bits or eight octets, and the remaining two octets can fit into one L2 frame. Therefore, one BEGIN frame and one END frame are used. The END frame for the normal time slot has 97-bit long data field. Of these, two octets or 16 bits are the left over L3 data and the remaining 81 bits are all zeros as the filler. The two L2 frames carrying the L3 message are shown below.

Field Name	BT	IDT	EHI	MSID	NL3M	L3LI	L3DATA	CRC
Bit-Setting	000	00	0	XX..XX	000	00001010	XXXXXX........XXXXXX	XX......XX
Field Length (bits)	3	2	1	20	3	8	64 (8 octets)	16

First L2 Frame – BEGIN Frame for the Normal Time Slot

Field Name	BT	RSVD	L3DATA	Filler	CRC
Bit-Setting	010	0	XX........XX	000.......0000	XX......XX
Field Length (bits)	3	1	16 (2 octets)	81	16

Second L2 Frame – END Frame for the Normal Time Slot

Case 3 – A single 22-octet L3 message is transmitted in multiple L2 frames for the normal time slot. Other parameters remain the same as Case 2. Since the L3 message is 22 octets long, three L2 frames are needed: one BEGIN-, one CONTINUE- and one END-frame as shown below.

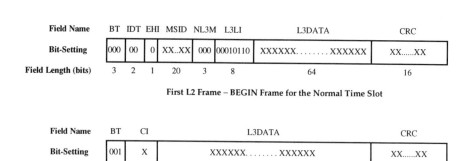

Field Name	BT	IDT	EHI	MSID	NL3M	L3LI	L3DATA	CRC
Bit-Setting	000	00	0	XX..XX	000	00010110	XXXXXX........XXXXXX	XX......XX
Field Length (bits)	3	2	1	20	3	8	64	16

First L2 Frame – BEGIN Frame for the Normal Time Slot

Field Name	BT	CI	L3DATA	CRC
Bit-Setting	001	X	XXXXXX........XXXXXX	XX......XX
Field Length (bits)	3	1	97	16

Second L2 Frame – CONTINUE Frame for the Normal Time Slot

Field Name	BT	RSVD	L3DATA	Filler	CRC
Bit-Setting	010	0	XX........XX	000.......0000	XX......XX
Field Length (bits)	3	1	15	82	16

Third L2 Frame – END Frame for the Normal Time Slot

Case 4 – Two 2-octet L3 messages are transmitted in a single L2 frames for the normal time slot. Other parameters remain the same as Case 2. One BEGIN-frame is used as shown below.

Field Name	BT	IDT	EHI	MSID	NL3M	L3LI1	L3LI2	L3DATA	Filler	CRC
Bit-Setting	000	00	0	XX..XX	010	00010110	00010110	X...X	00...00	XX......XX
Field Length (bits)	3	2	1	20	3	8	8	32	32	

BEGIN Frame for the Normal Time Slot

Now we discuss the SPACH ARQ STATUS frame shown in Figure 3.25. The BT-field is set to 100 for this frame. The CRC-field is same as that with the other frames. The RSVD-field is two bits long instead unlike that in the END frame. The following fields are new:

- *Partial Echo Assigned (PEA):* The field length is seven bits. This field is used by the MS during an ARQ mode of transaction.

■ *Frame Number Map (FRNO MAP):*   Receive status of the ARQ mode
transaction.

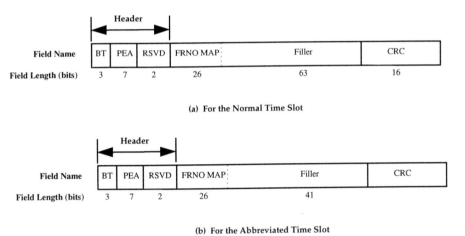

(a) For the Normal Time Slot

(b) For the Abbreviated Time Slot

Figure 3.25   General format of the RACH Layer 2 SPACH ARQ STATUS frame

The FDCCH Protocols

The FDCCH includes the BCCH, SPACH, SCF and Reserved channels.   The
protocols on these logical channels are discussed next.

Since the SCF channels is a fixed field of every FDCCH time slot, the SCF protocol
format is discussed as part of the discussion on the time slot format.   Here, the
protocols of the BCCH and SPACH channels are discussed first.

The Layer 2 frames for the BCCH and SPACH carry Layer 3 messages used on those
logical channels.   In addition, the SPACH also carries Layer 2 messages supporting
the ARQ protocol.   The Layer 2 frames for both the BCCH and the SPACH are
carried in the DATA-field of the FDCCH time slot.   As discussed earlier, a Layer 2
frame is first appended with five bits of the tail (T)-field and is encoded to yield a bit
stream twice the combined length of the Layer 2 frame and the T-field.   This bit
stream after the encoding must fit in the DATA-field of the FDCCH time slot.
Unlike the RDCCH for which the normal and abbreviated time slot formats are used,
the FDCCH employees only one time slot format in which the length of the DATA-

field is 260 bits. Since the encoding process doubles the number of bits, the length of the Layer 2 (L2) frame is calculated for the BCCH and the SPACH as follows:

(L2 frame length + 5 T-field bits) x 2 = DATA-field length

or

L2 frame length = DATA-field length/2 -5 = (260/2) - 5

= 125 bits

We first discuss the BCCH protocol and then the SPACH protocol.

The BCCH Protocol.   The BCCH has three sub-channels: the F-BCCH, the E-BCCH and the S-BCCH. The protocol for the last sub-channel has not yet been defined and is left for the further study. The protocols for the first two sub-channels are discussed here.

Only two types of frames are used for both sub-channels: the BEGIN frame and the CONTINUE frame. No END frame is used; the CONTINUE frame in this case serves the similar purposes of both the CONTINUE and END frames of the RACH.

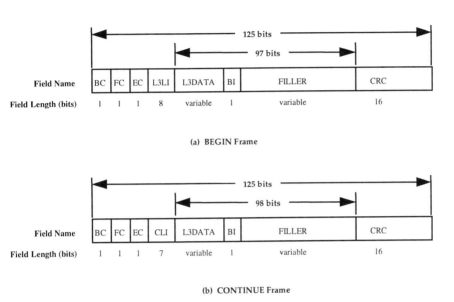

(a) BEGIN Frame

(b) CONTINUE Frame

Figure 3.26 General format of the F-BCCH Layer 2 BEGIN and CONTINUE frames

The following explains the fields of the BEGIN AND continue frames of the F-BCCH:

■ *Begin/Continue (BC):* The BC-field is one bit long. Its bit-setting equal to 0 means that the current frame contains the beginning of a Layer 3 message; 1 means that it contains a continuation (or the end) of the Layer 3 message.

■ *F-BCCH Change (FC):* The FC-field is one bit long. The bit in this field toggles between 0 and 1 whenever the information content of the F-BCCH changes beginning with the current hyperframe which contains the superframe carrying the F-BCCH.

■ *E-BCCH Change (EC):* The EC-field is one bit long. The bit in this field toggles between 0 and 1 whenever the information content of the E-BCCH changes beginning with the current hyperframe which contains the superframe carrying the E-BCCH.

■ *Layer 3 Message Length Indicator (L3LI):* See the same discussion for the RACH.

■ *Layer 3 Data (L3DATA):* See the same discussion for the RACH.

- *Begin Indicator (BI):* This field is one bit long. If its bit-setting is equal to 0, the remaining portion of the current frame after this field does not contain the beginning portion of a new Layer 3 message, and, therefore, is filled with the filler, which are all zeros. If its bit-setting is equal to 1, a new Layer 3 message portion begins right after this field.

- *Continuation Length Indicator (CLI):* The length in bits of the portion of the Layer 3 message continued in the current frame. If the current CONTINUE frame is the last frame and the length indicated by the CLI-field is not enough to completely fill the L3DATA-field, the remainder is filled by one BI-field bit set to 0 and the all-zero FILLER field.

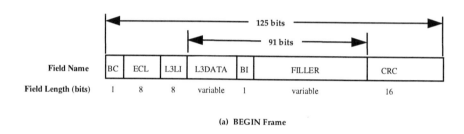

(a) BEGIN Frame

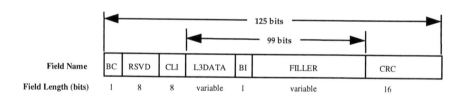

(b) CONTINUE Frame

Figure 3.27 General format of the E-BCCH Layer 2 BEGIN and CONTINUE frames

- *E-BCCH Cycle Length (ECL):* The field length is eight bits. This field indicates the total number of L2 frames required for the current E-BCCH cycle.

- *Reserved (RSVD):* This field contains one bit position, which is set to 0.

The SPACH Protocol

The SPACH includes three sub-channels: The PCH, the ARCH and the SMSCH. The first sub-channel operates in the unacknowledged mode only and the last two operate in either in a unacknowledged mode or an acknowledged mode. Two classes of frames are defined for the SPACH: one for the unacknowledged mode operation, and the other for the acknowledged mode operation. We will focus on the latter.

Two types of L2 frames are used for the ARQ protocol: the ARQ Mode BEGIN frame and the ARQ Mode CONTINUE frame. They serve similar functions as those of the BEGIN and CONTINUE frames discussed earlier, i.e., transporting Layer 3 messages, except that the qualifier "ARQ MODE" indicates that these L2 frames are used in the ARQ protocol. The header field of the SPACH L2 frame in general may contain up to three different headers: Header A, Header B, and Extension Header. The L2 frames used for the ARQ mode operation, i.e., the ARQ Mode BEGIN and ARQ Mode CONTINUE frames always include Header A and Header B. The ARQ Mode CONTINUE frame never includes the Extension Header; and the ARQ Mode BEGIN frame may or may not include the Extension Header. (Note that other SPACH frames without ARQ mode operation may include only Header A.) The ARQ Mode CONTINUE frame may serve as the end frame if it is the last L2 frame of the L3 message. In this case, the remaining portion of the L3DATA field of the frame, if any, is filled by all zeros as the filler. The two types of the ARQ Mode BEGIN frame and the ARQ Mode CONTINUE frame are shown below:

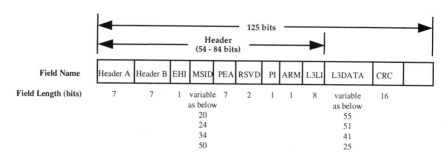

(a) ARQ Mode BEGIN Frames without the Extension Header

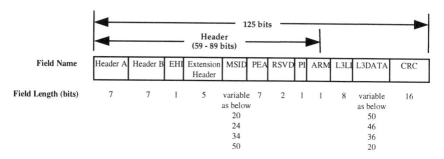

(b) ARQ Mode BEGIN Frames with the Extension Header

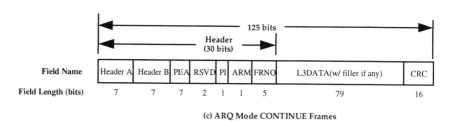

(c) ARQ Mode CONTINUE Frames

Figure 3.28 The general format of the ARQ Mode BEGIN and ARQ Mode CONTINUE frames for the SPACH ARQ mode protocol

Header A, Header B and Extension Header are further divided into the following sub-fields:

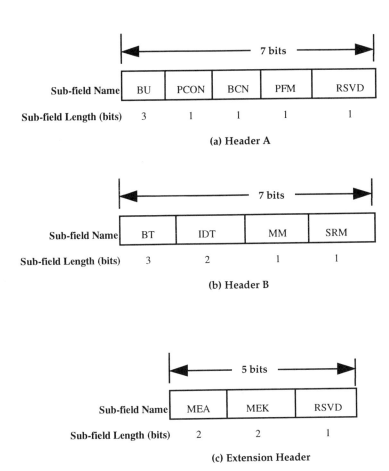

Figure 3.29 The three types of header sub-fields of the header of the SPACH L2 frames

The following are the fields and sub-fields of the ARQ mode protocol frames for the SPACH:

- *Header A:* This header field is present in all SPACH L2 frames including those without the ARQ protocols. The presence or absence of the next header field, Header B, is determined in the Header A field. The total length of this field is seven bits and consists of the following subfields:

- *Burst Usage (BU):* This sub-field is three bits long and allows eight different possibilities for the use of an L2 frame. The bit-settings of this sub-field correspond to the following frame: 000 = null; 001 = Hard Penta Page (20-bit MSID); 010 = Reserved; 011 = ARCH burst; 100 = SMSCH burst; 101 = Hard Triple Page (34-bit MSID); 110 = Hard Quadruple Page (20- or 24-bit MSID); 111 = Page burst. Since the ARQ protocol is supported by the ARCH sub-channel and the SMSCH sub-channel and not by the PCH sub-channel of the SPACH, this field must be either 011 (if it is for the ARCH) or 100 (if it is for the SMSCH) in the ARQ mode frames.

- *PCH Continuation (PCON):* This sub-field is one bit long and indicates PCH continuation if set to 1 and no PCH continuation if set to 0.

- *BCCH Change Notification (BCN):* This sub-field is one bit long and toggles between 0 and 1 whenever there is a change in the F-BCCH or the E-BCCH information.

- *Paging Frame Modifier (PFM):* This sub-field is one bit long.

■ *Header B:* This header field is present in all SPACH L2 frames including those without the ARQ protocols. The presence or absence of this header field is determined by the bit-settings of the BU-sub-field of the Header A field. The Header B field is present for the following three uses of the frame: BU-subfield of the Header A field set to 011 (ARCH frame), 100 (SMSCH frame) and 111 (PCH frame); for the other five bit-settings of the BU-subfield, no Header B field is present. The total length of the Header B field is seven bits and consists of the following subfield:

- *Burst Type (BT):* This sub-field is three bits long and allows eight different frame types. The bit-settings in this subfield are 101 for the ARQ Mode BEGIN frame and 110 for the ARQ Mode CONTINUE frame. The other six bit-settings apply to the frames without the ARQ protocol.

- *Identify Type (IDT):* This subfield is two bits long and allows four different lengths of the MS identity MSID)/user identify (USID): 20 bits (00), 24 bits (01), 34 bits (10) and 50 bits (11).

- *Message Mapping (MM)*

– *SPACH Response Mode (SRM):* Since the logical channels are unidirectional, in order to implement the ARQ protocol, the MS must use the reverse logical channel to respond to the SPACH message. The SRM-subfield is one bit long and indicates how the MS must access the reverse channel, the RACH, to respond to the SPACH message. If this field is set to 0, the MS must attempt to access the RACH on a contention basis; is set to 1, on a reservation basis.

■ Extension Header Indicator (EHI): This field is one bit long and indicates whether the next field, the Extension Header field, is present (1) or absent (0).

■ *Extension Header:* If the EHI-field is set to 1, this header field is present. This field contains information necessary in identifying the message encryption algorithm and the message encryption key used to encrypt the Layer 3 information payload. This field is five bits long and consists of the following subfields:

– *Message Encryption Algorithm (MEA)*

– *Message Encryption Key (MEK)*

■ *Partial Echo Assigned (PEA):* This field is seven bits long and is used in the ARP protocol operation, which will be discussed later.

■ *Polling Indicator (PI):* This field is one bit long and indicates whether or not the BS is requesting the MS to send the ARQ STATUS frame on the reverse channel, the RACH. If this field is set to 1, the MS must send the ARQ STATUS frame; if set to 0, the MS is not required to send the ARQ STATUS frame. More will be discussed on this when we discuss the ARQ protocol later.

■ *ARQ Response Mode (ARM):* This field is one bit long and indicates how the MS must respond on the RACH when the PI-field is set to 1. If the ARM-field is set to 1, the ARQ STATUS frame must be sent on the RACH on a reservation basis; if set to 0, on a contention basis. More will be discussed on this when we discuss the ARQ protocol later.

The SPACH ARQ Protocol

The Automatic Repeat Request (ARQ) protocol is used as an error control technique. The ARQ protocol requires positive acknowledgments of the L2 frames sent and automatic retransmission of un acknowledged frames. The ARQ protocol is used for the two of the SPACH subchannels – the ARCH and the SMSCH. The PCH subchannel does not operate in the acknowledged mode and, therefore, does not use the ARQ protocol.

The ARQ protocol is implemented by using the ARQ MODE BEGIN and the ARQ MODE CONTINUE frames on the forward direction, the SPACH, and the SPACH ARQ STATUS frame on the reverse direction, the RACH. This exchange of protocol frames is illustrated in Figure 3.30.

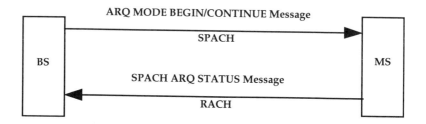

Figure 3.30 The ARQ Mode

When the ARQ mode of operation is not in effect, the MS is said to be in the ARQ idle state. When the MS receives an ARQ MODE BEGIN frame, it first examines the MSID field in the message. If the content of the MSID field matches its own MSID, the MS enters the ARQ active state and starts the MS ARQ procedure based on the parameter values read from the ARQ MODE BEGIN message as follows. The PEA value uniquely identifies the current instance of the ARQ transaction. the MS stores this value. The L3LI indicates the length of the Layer 3 message transported by the current ARQ mode of operation, and allows the MS to determine the number of the subsequent ARQ MODE CONTINUE frames necessary to complete the Layer 3 message.

Recall that the ARQ protocol uses the SPACH ARQ STATUS frame on the reverse direction, the RACH. The FRNO MAP field of the SPACH ARQ STATUS frame is the means of communicating to the BS whether or not the frames have been received correctly by the MS. This field is 26 bits long and must be initialized by setting all bit positions to 0. The 26 bit positions of the FRNO MAP field of the SPACH ARQ STATUS frame correspond to 26 ARQ MODE CONTINUE frames. The ARQ

MODE CONTINUE frame with the FRNO-field bit set to 0 is mapped to the left most bit position of the FRNO MAP field of the SPACH ARQ STATUS frame and the ARQ MODE CONTINUE frame with the FRNO-field bit set to 25, to the right most bit position of the FRNO MAP field of the SPACH ARQ STATUS frame. This FRNO mapping is illustrated in Figure 3.31.

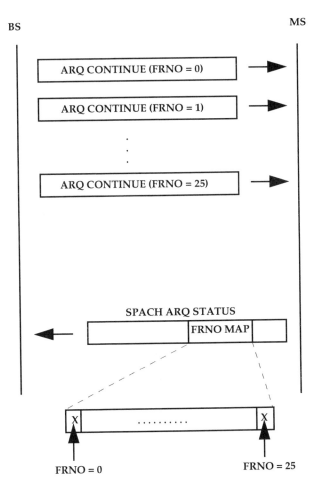

Figure 3.31 The ARQ protocol

If the ARQ CONTINUE frame with the FRNO j is correctly received, the $(j+1)^{th}$ bit position from the left in the FRNO MAP field in the SPACH ARQ STATUS frame is set to 1, and if the frame is not correctly received, the same bit position is set to 0. The MS stores the Layer 3 message segment contained in the L3DATA field of the ARQ MODE BEGIN or CONTINUE message currently received.

The MS responds to the BS with the ARQ STATUS frame based on the polling by the BS. This allows the BS to control the amount of traffic it can process at any given time. The BS polls the MS to send the ARQ STATUS frame by setting the PI bit in the ARQ MODE BEGIN/CONTINUE frame to 1. If PI = 1 and ARM = 1, the ARQ STATUS frame is sent on a reservation basis, and, if PI = 1 and ARM = 0, sent on the contention basis. Whether the ARQ STATUS frame is successfully sent or not is determined by reading the R/N bits of the SCF channel. If all five of the R/N bit positions are set to 1, the ARQ STATUS frame is successfully received; otherwise, it is not.

If the ARQ STATUS frame is successfully received, the MS proceeds as follows. If one or more ARQ MODE CONTINUE frames are pending, the ARQ mode is continued as described further below; if no ARQ MODE CONTINUE frames are pending, the MS terminates the ARQ mode and enters the ARQ idle state. If the ARQ STATUS frame is not successfully sent, the MS terminates the ARQ mode and enters the ARQ idle state. If PI = 0, the MS proceeds as follows. If one or more ARQ MODE CONTINUE frames are pending, the ARQ mode is continued as described further below; if no ARQ MODE CONTINUE frames are pending, the MS terminates the ARQ mode and enters the ARQ idle state.

Suppose that the ARQ mode is continuing. If an ARQ MODE CONTINUE frame with matching PEA value is not received within 32 TDMA blocks after a previously received ARQ MODE BEGIN/CONTINUE frame or within 32 TDMA blocks after successfully sending a SPACH ARQ STATUS frame, the MS sends an unsolicited ARQ STATUS frame to the BS. If this unsolicited ARQ STATUS frame is not sent successfully, the MS terminates the ARQ mode and enters the ARQ idle state. If an ARQ MODE CONTINUE frame is not received within 32 TDMA blocks after successfully sending an unsolicited ARQ STATUS frame, the MS terminates the ARQ mode and enters the ARQ idle state. iF AN ARQ MODE BEGIN frame with an MSID matching its own is received while executing the current ARQ mode procedure, the MS terminates the current ARQ mode and starts a new ARQ mode. If an ARQ MODE CONTINUE frame with matching PEA value is received within 32 TDMA blocks, the MS proceeds as described earlier.

3.3 THE DIGITAL TRAFFIC CHANNEL

3.3.1 The Physical Channel Structure

The physical channel structure of the Digital Traffic Channel is similar to that of the DCCH. The frame duration is 40 ms; the frame length is 1944 bits; each frame contains six time slots; and each time slot contains 324 bit positions. The time slot formats for the forward and reverse traffic channels are shown in Figure 3.32.

Field Name	G	R	DATA	SYNC	DATA	SACCH	CDVCC	DATA
Bit Position	1-6	7-12	13-28	29 - 56	57 - 178	179 - 190	191 - 202	203 - 324
Field Length (bits)	6	6	16	28	122	12	12	122

(a) Reverse DTC Time Slot

Field Name	SYNC	SACCH	DATA	CDVCC	DATA	RSVD	CDL
Bit Position	1- 28	29 - 40	41 - 170	171 - 182	183 - 312	313	314-324
Field Length (bits)	28	12	130	12	130	1	11

(b) Forward DTC Time Slot

Legends

G	- Guard Time	SACCH	-	Slow Associated Control Channel
R	- Ramp Time	CDVCC	-	Coded Digital Verification Color Code
DATA	- User information or FACCH	CDL	-	Coded Digital Control Channel Locator
SYNC	- Synchronization			

Figure 3.32 The time slot formats of the DTC

The Digital Traffic Channel (DTC) contains three subchannels: the user data, carried in the DATA-field of the time slot; the Slow Associated Control Channel (SACCH) carried in the SACCH-field of the time slot; and the Fast Associated Control Channel (FACCH) carried in the DATA-field of the time slot. The DATA-field can be used to carry either the user data or the FACCH at one time but not both simultaneously.

Each full-rate DTC uses two time slots per frame and each half-rate DTC, one time slot per frame. In both the reverse and forward DTC (RDTC), 260 bits per time slot are allocated for the user data or the FACCH and 12 bits, for the SACCH. Hence, the bit transmission rates for the user data/FACCH and the SACCH for the full-rate and half-rate DTC's are as shown by Table 3.2.

DTC Subchannel	Number of Bits per Frame	Bit Transmission Rate (kb/s)
Full-Rate DTC		
User Data/FACCH	520	13.0
SACCH	24	0.6
Total	544	13.6
Half-Rate DTC		
User Data/FACCH	260	6.5
SACCH	12	0.3
Total	272	6.8

Table 3.2   The transmission speeds of the subchannels of the DTC

3.3.2 The Power Output Characteristics

The carrier-off condition is defined as a power output at the transmitting antenna connector not exceeding -60 dBm. When commanded to turn off the carrier, the MS transmit power must fall to this level or below within three symbol periods, which is about 123 μsec.

The nominal Effective Radiated Power (ERP), which is the average burst power in digital mode, for each class of MS transmitter is as follows:

Class II	0 dBW (1.0 Watts)
Class IV	-2dBW (0.6 Watts)
Class III, V - VIII	Reserved

All MS transmitters must be capable of reducing or increasing power on command from the BS. The MS power levels (PL's), the Digital Mobile Attenuation Code (DMAC) and the corresponding nominal ERP's for the MS power classes are shown in Table 3.3.

Mobile Station Power Level (PL)	*DMAC*	*Nominal ERP in dBW for the MS*						
		II	*III*	*IV*	*V*	*VI*	*VII*	*VIII*
0	0000	0	.	-2	.	.	.	.
1	0001	0	.	-2	.	.	.	.
2	0010	-2	.	-2	.	.	.	.
3	0011	-6	.	-6	.	.	.	.
4	0100	-10	.	-10	.	.	.	.
5	0101	-14	.	-14	.	.	.	.
6	0110	-18	.	-18	.	.	.	.
7	0111	-22	.	-22	.	.	.	.
8	1000	-26	.	-26	.	.	.	.
9	1001	-30	.	-30	.	.	.	.
10	1010	-34	.	-34	.	.	.	.

Table 3.3 The Mobile Station (MS) nominal power levels

3.3.3 Speech Coding

The current subsection discusses the speech coding; and the next subsection will deal with the channel coding.

There are three broad classes of speech coding techniques: waveform coding, linear predictive coding, and hybrid coding. In the waveform coding method, the analog speech waveform is sampled and the sampled speech amplitudes are coded in binary bits so that the original speech waveform may be reproduced at the receiving end as faithfully as possible. In the linear predictive coding, the human vocal tract is modeled by an all-pole digital filter, i.e., a filter with poles and no zeros. The speech signal is then predicted using the model whose key parameters are periodically updated as the excitation signal, which is a function of the individual's speech, varies. The key parameters of the model are coded by digital bits for transmission. A hybrid codec combines the two methods.

This system uses a hybrid coding, which is a variation of the Code Excited Linear Predictive Coding (CELP) method called the Vector-Sum Excited Linear Predictive Coding (VSELP). In this method, the analog speech signal is first sampled, quantized and coded into binary bits just as in the typical waveform codec such as the Pulse Code Modulation (PCM). The sampling rate of the full-rate speech coder is 8000 samples per second, or 8 kHz. Each sample is represented by eight bits. The standard μ-law companding is used. The result of this A/D conversion is a 64 kb/s bit stream in the PCM format. This initial bit stream is used for further digital signal processing for the vocal tract modeling.

For the purpose of this processing, speech samples are grouped into frames. One frame is 20 ms in duration. Since the sampling rate is 8 kHz, one frame contains 160 speech samples. Each frame is further divided into four subframes. Each subframe is 5 ms in duration and contains 40 speech samples. The 160 samples of speech amplitudes of the frame are represented by 159 bits. These 159 bits are allocated as follows:

Short term filter coefficients	38 bits
Frame energy	5 bits
Lag	28 bits
Codewords	56 bits
Gain parameters	32 bits
Total per frame	159 bits

Since 159 bits representing the speech are transmitted in every frame duration, which is 20 ms, the basic data rate of the codec is 7.95 kb/s. This also implies that 8000 samples per second of information are represented by 7950 bits per second, amounting to slightly less than one bit to represent one sample of speech amplitude. As a comparison, the Pulse Code Modulation (PCM) codec uses eight bits to code one speech sample at the sampling rate of 8 kHz yielding 64 kb/s data rate; similarly, the Adaptive Differential Pulse Code Modulation (ADPCM) uses four bits to code one speech sample at the sampling rate of 8 kHz yielding 32 kb/s data rate. Figure 3.33 illustrates the speech coding process used in this system.

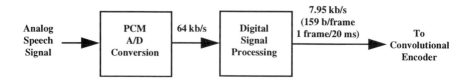

Figure 3.33 Speech coding process

3.3.4 Channel Coding

The output of the speech encoder is the 159 bits per 20-ms frame. This bit stream is inputted into the channel coding process.

Data Classes

For the purpose of the channel coding, the 159 bits per frame is first divided into two classes:

- *Class 1 bits.* There are 77 class 1 bits per frame. These are more important bits and are subject to error protection.

- *Class 2 bits.* The remaining 82 bits per frame are class 2 bits. These bits are transmitted without error protection. However, they are included in the interleaving process.

Out of the 77 class 1 bits, 12 bits are selected as the most perceptually significant bits. Table 3.4 shows the assignments of the 159 bits to various parameters of the speech codec and their breakdown of class 1 and class 2 bits and the 12 most perceptually significant bits.

Parameter	Total Number of Bits	Number of Class 1 Bits	Number of Class 2 Bits	Number of Most Perceptually Significant Bits
Frame energy	5	4	1	3
Reflection coefficients	38	13	25	9
Subframe lags	28	28	0	0
1st codebook code	28	0	28	0
2nd codebook code	28	0	28	0
Gain parameters	32	32	0	0
Total	159	77	82	12

Table 3.4 The bit assignments of the speech codec parameters

CRC

The 12 most perceptually significant bits are protected by a seven-bit CRC which is computed using a polynomial in which the 12 bits appear as coefficients as follows. First, the input polynomial, $a(X)$, to the CRC computation is an eleventh order polynomial in which the 12 most perceptually significant bits denoted by m_0, m_1, m_2, ..., m_{11} appear as the coefficients as follows:

$$a(X) = m_0 + m_1X + m_2X^2 + m_3X^3 + \ldots\ldots + m_{10}X^{10} + m_{11}X^{11}.$$

The generator polynomial for the CRC, $g(X)$, is a seventh order polynomial as follows:

$$g(X) = 1 + X + X^2 + X^4 + X^5 + X^7.$$

The input polynomial is multiplied by X^7 and divided by the generator polynomial as follows:

$$a(X)\ X^7/g(X) = q(X) + b(X)/g(X).$$

The quotient $q(X)$ is discarded. The remainder polynomial $b(X)$ is a sixth order polynomial of the form:

$$b(X) = r_0 + r_1X + r_2X^2 + r_3X^3 + r_4X^4 + r_5X^5 + r_6X^6.$$

The seven coefficients of $b(X)$, r_0, r_1, r_2, r_3, r_4, r_5 and r_6, are included in the convolutional encoding process discussed next.

Convolutional Encoding

The input to the convolutional encoder is a 89-bit digital bits stream composed of the 77 class 1 bits, the seven bits of CRC and five tail bits. These five tail bits are all zeros and are appended to make the number of bits transmitted for each speech frame fit evenly into the time slot DATA-field. The bit positions of this 89 bits are shown in Figure 3.34. The mapping of the seven CRC bits, r_0, r_1, r_2, r_3, r_4, r_5 and r_6, into the specific bit positions in this bit stream is also shown in the figure.

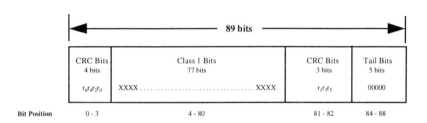

Figure 3.34 The input to the convolutional encoder

For the convolutional encoding, the code rate, R, is 1/2 and the memory order, m, is five. With R = 1/2, for each input bit, two bits are produced and the 89 input bits result in 178 bits at the output of the encoder. The 82 class 2 bits, which are not encoded, are added to these 178 bits to make a total of 260 bits for a 20 ms speech frame.

Interleaving

Interleaving is performed to minimize the number of burst errors resulting from the Raleigh fading. After the encoding, 260 bits represent one speech frame, which fit evenly into the DATA-field of one time slot. The interleaving is performed over two time slots. This means the that the bits representing two speech frames, which amount to 520 bits, are interleaved and the resulting 520 interleaved bits are put into two time slots. Therefore, each time slot contains portions of the two adjacent speech frames. The bits are entered into an interleaving array column-wise and read out row-wise according to certain order of mixing the data.

3.3.5 Modulation

The speech data, after encoding and interleaving, are then put into time slots. Each time slot contains 324 bits. Six time slots make up one TDMA frame; one TDMA frame contains 1944 bits; and the duration of one TDMA frame is 40 ms. This yields the bit transmission rate of 48.6 kb/s. This digital bit stream is used to modulate the carrier. This system uses the $\pi/4$ Differentially-encoded Quadrature Phase Shift Keying (DQPSK) modulation.

3.3.6 Security and Identification

The following information is stored in the User Identity Module (UIM), which is implemented as either a logical entity within the personal station or as a separate module:

- *Mobile Identification Number or International Mobile Subscriber Identify (IMSI)*
- *Home System Identification (HSI)*
- *Shared Secret Data (SSD)*
- *Random Challenge Memory (RAND)*
- *Call History Parameter*
- *A-Key*
- *Updating the SSD*
- *CAVE Algorithm*

In addition, the following information is stored in the personal station:

- *Electronic Serial Number (ESN)*
- *Station Class Mark*
- *Discontinuous-Transmission*

Mobile Identification Number (MIN)

The Mobile Identification Number (MIN) is 34 bits long and is derived from the MS's 10-digit directory number, i.e., three digits for the Numbering Plan Area (NPA), three digits for the office NXX code, and four digits for the individual telephone number, XXXX. The three NPA digits are converted into 10 binary bits, which are denoted as MIN2 by the following algorithm:

1. Represent the three digit field as $D_1D_2D_3$ with the digit 0 having the value 10.
2. Compute $100\,D_1 + 10D_2 + D_3 - 111$

3. Convert the result in Step 2 to a standard decimal-to-binary conversion as follows:

Decimal Number	Binary Number
1	0000000001
2	0000000010
3	0000000011
4	0000000100
.	.
.	.
.	.
998	1111100110
999	1111100111

The three NXX digits are converted to 10 binary bits by the same algorithm as that described above. These 10 bits are mapped into the 10 most significant bits MIN1. Finally, the four XXXX bits are converted to 14 binary bits by the following algorithm and are mapped into the 14 least significant bits of MIN1$_p$:

1. The last three digits of XXXX are converted into 10 binary bits by the same algorithm as that described above.

2. The thousands digit, i.e., the first digit of XXXX, is converted into four bits by a Binary-Coded-Decimal (BCD) conversion as follows:

Thousand Digit	Binary Sequence
1	0001
2	0010
3	0011
4	0100
5	0101
6	0110
7	0111
8	1000
9	1001
0	1010

The 34-bit MIN obtained by the above algorithms is shown in Figure 3.35.

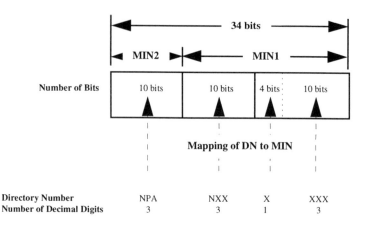

Figure 3.35 The mapping of the Directory Number (DN) to the Mobile Identification Number (MIN)

Example

A 10-digit Directory Number (DN) 908-758-2442 is converted to the MIN as follows.

First, the NPA 908 is converted to 10-bit MIN2 as follows:

$$D_1 = 9; D_2 = 10; D_3 = 8$$

$$100D_1 + 10D_2 + D_3 - 111 = 100(9) + 10(10) + 8 - 111 = 897$$

The decimal number 897 in the binary number system is 1110000001. Hence, MIN2 is 1110000001.

Next, the NXX 758 is converted to the 10 most significant bits of MIN1 as follows:

$$D_1 = 7; D_2 = 5; D_3 = 8$$

$$100D_1 + 10D_2 + D_3 - 111 = 100(7) + 10(5) + 8 - 111 = 647$$

The decimal number 647 in the binary number system is 1010000111. The next four most significant bits of MIN1 are derived from the first digit of XXXX, which is 2, which in the BCD coding is 0010. Finally, the 10 least significant bits of MIN1 are derived as follows:

$$D_1 = 4; D_2 = 4; D_3 = 2$$

$$100D_1 + 10D_2 + D_3 - 111 = 100(4) + 10(4) + 2 - 111 = 331$$

The decimal number 331 in the binary number system is 0101001011. Hence, MIN1 is 10100001100100101001011. Concatenating MIN2 and MIN1, the MIN for this DN is 1110000001101000011100100101001011.

Electronic Serial Number (ESN)

The ESN is set at the factory. The ESN is a 32-bit binary code, which uniquely identifies the MS and consists of eight bits of the Manufacturer's (MFR) code, six bits of reserved space and 18 bits of a serial number as follows:

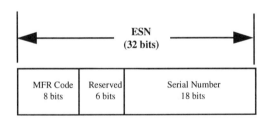

Station Class Mark (SCM)

The Station Class Mark (SCM) is a five-bit code to indicate the MS's power class and the mode of transmission. The first, second and fifth bits are used to indicate the power class; the third bit is used to indicate the transmission mode; and the fourth bit is not used.

The power classes and the SCM bit settings are as follows:

Power Class	SCM Bit-Setting
Class I	0XX00
Class II	0XX01
Class III	0XX10
Class IV	0XX11
Class V	1XX00
ClassVI	1XX01
Class VII	1XX10
Class VIII	1XX11

The transmission mode and the SCM bit-settings are as follows:

Transmission	SCM Bit-Setting
Continuous	XX0XX
Discontinuous	XX1XX

Home System Identification

A 15-bit System Identification Indicator (SID) is stored in the MS to identify the MS's home system. The SID consists of two bits for the international (INTL) code and 13 bits for the system number as follows:

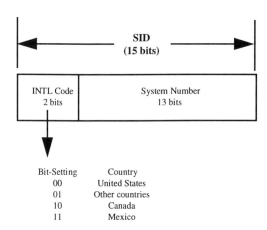

The 13-bit serial number is assigned to each United States system by the Federal Communications Commission (FCC).

Authentication

The term authentication refers to the process by which the BS confirms the identity of the MS. In the authentication process, the BS and the MS exchange certain information. If, based on this exchange of information, it can be shown that the BS and the MS possess identical sets of the Shared Secret Data (SSD), the MS is authenticated; otherwise, the authentication results in a failure.

The authentication procedure is depicted in Figure 3.36.

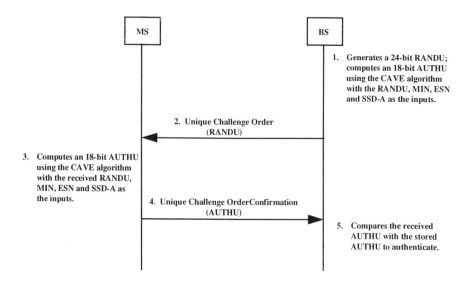

Figure 3.36   The Challenge-Response process of authentication

1. The BS generates a 24-bit random pattern referred to as RANDU and computes a 18-bit parameter AUTHU using the CAVE algorithm and the following parameters as the inputs to the CAVE algorithm: the RANDU, 8-bit MIN2, 24-bit MIN1, 32-bit ESN and 64-bit Shared Secret Data-A (SSD-A).

2. The BS sends the same RANDU to the MS via the Fast Associated Control Channel (FACCH) of the Forward Digital Traffic Channel (FDTC) in a *Unique Challenge Order Message*.

3. Upon receiving the 24-bit random pattern RANDU from the BS, the MS computes a 18-bit AUTHU by the same method as Step 1 using the RANDU received from the BS and the rest of the parameters, which are stored in the MS.

4. The MS sends the resulting 18-bit AUTHU to the BS via the FACCH of the Reverse Digital Traffic Channel (RDTC) in a *Unique Challenge Confirmation Message*.

5. The BS compares the AUTHU received from the MS and the AUTHU previously computed and stored by itself. If the two matches, the authentication is successful; otherwise, it is unsuccessful.

Figure 3.37 shows the inputs and output of the CAVE algorithm.

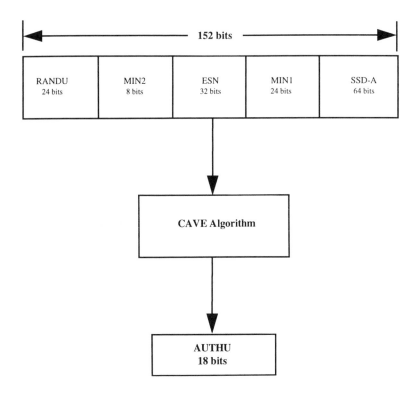

Figure 3.37 The inputs and the output of the CAVE algorithm

The Shared Secret Data (SSD) is a 128-bit pattern stored in the MS and readily available to the BS and is partitioned into two parts, 64-bit SSD-A and 64-bit SSD-B. SSD-A is used in the authentication process and SSD-B, in the voice privacy. Figure 3.38 shows the partition of SSD.

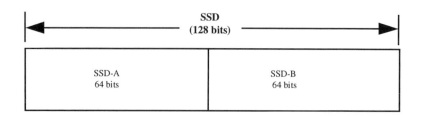

Figure 3.38   The contents of the Shared Secret Data (SSD)

Figure 3.39 shows the procedure of generating the SSD.

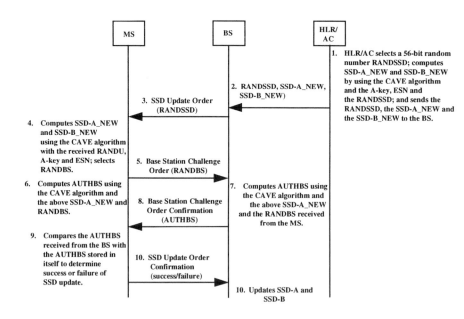

Figure 3.39   The procedure of generating the SSD

1. At the HLR/AC associated with the MS, a 56-bit random number referred to as RANDSSD is generated. Using this RANDSSD, A-key and ESN and the CAVE algorithm, SSD-A_NEW and SSD-B_NEW are computed. The A-key is a 64-bit code which is known only to the MS and its associated HLR/AC and is stored in the MS's permanent security and identification memory.

2. The RANDSSD, SSD-A_NEW and SSD-B_NEW are sent to the BS from the HLR/AC.

3. The BS sends an *SSD Update Order Message* to the MS via the FACCH of the FDCT. Included in that message is the RANDSSD received from the HLR/AC.

4. Upon receiving the *SSD Update Order Message*, the MS computes a 128-bit SSD using the CAVE algorithm and the following parameters as the inputs to the CAVE algorithm: the 56-bit RANDSSD, which is received from the BS, 32-bit ESN, and 64-bit A-key. Of the 128 bits computed as the output of the CAVE algorithm, the 64 most significant bits are assigned to SSD-A_NEW and the 64 least significant bits, to SSD-B_NEW.

5. The MS then selects a 32-bit random number, RANDBS, and sends it to the BS in a *Base Station Challenge Order Message* via the FACCH of the RDTC.

6. The MS computes a 18-bit AUTHBS using the CAVE algorithm and the following parameters as the inputs to the CAVE algorithm: the 34-bit RANDBS, which is sent to the BS, 32-bit ESN, 24-bit MIN1, and 64-SSD-A_NEW.

7. Upon receiving the *Base Station Challenge Order Message* from the MS, the BS computes an 18-bit AUTHBS by the same method as Step 6 using the RANDBS received from the MS and the rest of the parameters.

8. The BS sends a *Base Station Challenge Order Confirmation Message* to the MS via the FACCH of the FDCT. Included in that message is the AUTHBS computed in Step 7.

9. Upon receiving the *Base Station Challenge Order Confirmation Message*, the MS compares the AUTHBS contained in that message with the same stored in itself. If the two match, the MS replaces the contents of SSD-A and SSD-B with SSD-A_NEW and SSD-B_NEW, respectively, and sends an *SSD Order Confirmation Message* with SSD_UPDATE information element set to 1 to the BS via the FACCH of the RDCT. If the comparison results in a failure, the MS discards the SSD-A_NEW and SSD-B_NEW and sends an *SSD Order*

Confirmation Message with SSD_UPDATE information element set to 0 to the BS via the FACCH of the RDCT.

10. If the *SSD Order Confirmation Message* received from the MS indicates a success, the BS replaces the contents of SSD-A and SSD-B with the SSD-A_NEW and SSD-B_NEW received from the HLR/AC, respectively.

3.3.7 Supervision

Two types of supervisory functions are provided. The first type includes the traditional telephony supervisory functions of detecting "switch-hook" state. The second type is ensuring that adequate radio frequency signal quality is maintained.

Signaling Channels

Two subchannels of the DTC are used for the supervisory and control signaling: the Fast Associated Control Channel (FACCH) and the Slow Associated Control Channel (SACCH). The SACCH is a fixed 12-bit field of the time slot in both the RDTC and the FDTC and is present in all time slots along with the user DATA-field. Therefore, the SACCH is an out-of-band signaling channel, which may be used in parallel to the user data channel.

The FACCH and the user data channel occupy the same field in the time slots, i.e., the DATA-field. The 260-bit DATA-field is used to carry the user data or the FACCH but not at the same time. This means that the FACCH is an inband signaling channel. Certain messages may be sent over either the FACCH or the SACCH.

3.3.8 Signaling Format

The FACCH Word

The FACCH occupies 260-bit DATA-field of the DTC time slot. Each FACCH *word* is 65 bits long, which is convolutionally encoded with the code rate R = 1/4 to yield 260 bits. These 260 bits representing one FACCH word is put into the DATA-field of the DTC time slot. A FACCH message may span multiple FACCH words.

Figure 3.40 shows the format of the FACCH word.

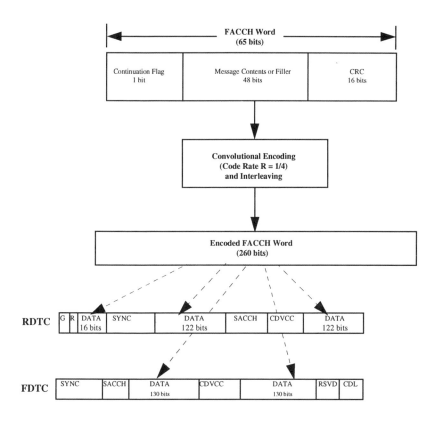

Figure 3.40 The FACCH word format and the mapping of the FACCH word into RDTC and FDTC time slots

The continuation flag equal to 0 indicates that the word is the first word of the message; 1, the subsequent word.

The SACCH Word

Figure 3.41 shows the SACCH word format.

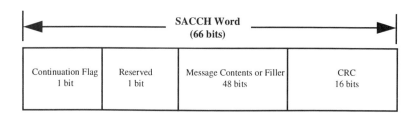

Figure 3.41 The word format of the SACCH

Each word is convolutionally encoded with the code rate R = 1/2, interleaved and transmitted in the SACCH field over 12 time slots.

The FACCH and SACCH Message Format

The same message format is used for the FACCH and the SACCH. Figure 3.42 shows the message format on these channels.

Protocol Discriminator	Message Type	Mandatory Fixed Information Elements	Mandatory variable Information Elements	Remaining Length	Optional Vartiable Information Elements
2 bits	8 bits			6 bits	

Figure 3.42   The FACCH and SACCH message format

The FACCH and SACCH ARQ Protocols

The ARQ mode of transmission is supported on the FACCH and the SACCH of the DTC.

4

THE STANDARD BASED ON THE NORTH AMERICAN HIGH-TIER CDMA SYSTEM

4.1 OVERVIEW

This wireless access system for the 1.9 GHz PCS is derived from the existing North American 900-MHz CDMA cellular standard referred to as the IS-95 standard[2]. The major characteristics of this system are highlighted below:

■ *Bandwidth.* Channel bandwidth is 1.25 MHz.

■ *Logical channel structure.* The Access Channel is used by the mobile station to initiate communication with the base station and to respond to paging channel messages. Each Access Channel frame contains 96 bits consisting of 88 information bits and eight encoder tail bits. The Reverse Traffic Channel is used for the transmission of user and signaling information to the base station during a call. The Paging Channel is used by the base station to transmit system overhead information and mobile station specific messages to the mobile station. The Forward Traffic Channel is used by the base station for the transmission of user and signaling information to a specific mobile station during a call.

■ *Information multiplexing.* Different types of information may be multiplexed in the I-Fields of the Traffic Channels.

- *Multiple access.* This system uses the CDMA method.

- *Authentication.* The authentication procedure used in this standard is essentially
  the same as that used in the North American TDMA cellular standard discussed
  earlier.

- *Handoff.* This standard provides the following three types of handoff: soft
  handoff, CDMA-to-CDMA hard handoff, and CDMA-to-Analog handoff. Soft
  handoff is a handoff in which the mobile station begins communications with a
  new base station without interrupting communications with the old base station.
  This type of handoff can only be used between CDMA channels having identical
  frequency assignments. CDMA-to-CDMA handoff is a handoff in which the
  mobile station is transitioned between disjoint sets of base stations, different
  frequency assignments, different band classes, or different frame offsets.
  CDMA-to-Analog handoff is a handoff in which the mobile station is directed
  from a Forward Traffic Channel to an analog voice channel of the 800 MHz
  system.

4.2   PHYSICAL CHANNELS

4.2.1 Method of Physical Channel Creation

In this system, physical channels are created by a combination of the Frequency
Division Duplexing (FDD), the Frequency Division Multiple Access (FDMA) and
the Code Division Multiple Access (CDMA) techniques as described below.

The Creation of the Reverse and
Forward Links by
the Frequency Division Duplexing (FDD)

First, two separate frequency bands are assigned for the forward and reverse
directions of transmission using the FDD technique. The FCC allocated 60 MHz of
spectrum from 1850 to 1910 MHz, the low band of the A-, B-, C-, D-, E-, and F-
blocks, and 60 MHz of spectrum from 1910 to 1990 MHz, the high band of the A-,
B-, C-, D-, E-, and F-blocks, for the licensed PCS operation. In this system, the low
60-MHz band, i.e., the 1810-1910 MHz band, is assigned for the reverse link and the

high 60-MHz band, i.e., the 1930-1990 MHz band, for the forward link. Figure 4.1 illustrates the creation of the reverse and forward links by the FDD for this system.

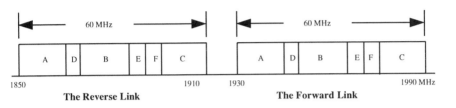

Figure 4.1 The frequency bands for the reverse and forward links created by the FDD technique

The Creation of the Frequency Channels by the Frequency Division Multiple Access (FDMA)

The 60-MHz-wide frequency band for each direction is then divided into smaller bands of frequency. These subbands of frequency are referred to as frequency channels, and are created by the FDMA technique. The center frequencies of two neighboring frequency channels are separated by 1.25 MHz. However, the bandwidth of each of these frequency channels is 1.23 MHz leaving a 20 kHz of gap between the two neighboring frequency channels.

These 1.23-MHz-wide frequency channels are created in two steps. The first step is to divide the 60-MHz-wide band into a large number of small frequency bands 30 kHz apart. These 50-kHz-wide bands are designated as the CDMA channels with numerical channel number identifications, and serve as building blocks of the 1.23-MHz-wide frequency channels. At this point, the term "CDMA channel" could be confusing because these are still the channels created by the FDMA and the CDMA technique has not yet been applied. Its meaning will become clearer after discussing the second step in creating the 1.23-MHz frequency channels next.

Once the 50-kHz-wide CDMA channels are created, the second step is to select appropriate channels from these CDMA channels and use their center frequencies as the center frequencies of the 1.23-MHz channels to be created. The neighboring 50-

khz channels on both sides of the selected CDMA channel are merged to create a larger bandwidth, i.e., 1.23 MHz. Since the center frequency of the selected CDMA channel becomes the center frequency of the 1.23-MHz frequency channel, the term "CDMA channel" may be interpreted as the "channel that gives the center frequency for the 1.23-MHz frequency channel." This two-step process of creating the 50-kHz-wide CDMA channels and the 1.23-MHz-wide frequency channels is discussed below.

First, the creation of the 50-kHz-wide CDMA channels is discussed. The 60-MHz-wide band from 1810 MHz to 1910 MHz and that from 1930 to 1990 MHz are divided into the 50-kHz-wide CDMA channels with the center frequencies, f_N, given by the following equations:

Reverse link

$$f_N = 1850.00 + 0.050 \text{ MHz}, \qquad N = 0, 1, 2, \ldots, 1199$$

Forward link

$$f_N = 1930.00 + 0.050 \text{ MHz}, \qquad N = 0, 1, 2, \ldots, 1199$$

where N denotes the channel number.

As an illustration of this process, Figure 4.2 shows the CDMA channel designations for the A-block of frequency. Similar illustrations are possible for the other frequency blocks. Since the A-block has a total of 15 MHz of bandwidth in each direction, a total of 300 channels are created (Channels 0-299) in each direction, i.e., 300 two-way channels. Similarly, the B- and C-blocks are both 15-MHz wide in each direction and have 300 two-way channels. The D-, E- and F-blocks are 5-MHz wide in each direction, and have 100 two-way channels.

Next, the creation of the 1.23-MHz-wide frequency channels from the CDMA channels is discussed.

Certain of the 50-kHz channels are selected and their center frequencies are made the center frequencies of the 1.23-MHz frequency channels to be created. For example, for the A-block channels, as shown in Figure 4.2, Channels 0 - 24 are not used, Channels 276-299 are used conditionally, and only Channels 25-275 are valid candidates for the center frequencies for the 1.23-MHz channels. The conditionally valid channels are permitted if the adjacent block is allocated to the same service provider (in this example if the D-block, which is adjacent to the A-block, also belongs to the same A-block operator).

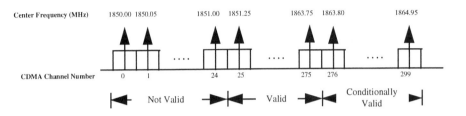

The Reverse Link

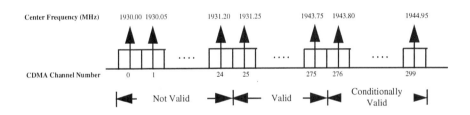

The Forward Link

Figure 4.2   The designations of the CDMA channel for the A-block

The number of 1.23-MHz channels that can be created in the six frequency blocks and the preferred center frequencies of the 1.23-MHz frequency channels are shown in Table 4.1.

Freq. Blocks	No. of Freq. Chan	CDMA Channel Numbers Whose Center Frequencies are Preferred as the Center Frequencies of the 1.23-MHz Frequency Channels
A	11	25, 50, 75, 100, 125, 150, 175, 200, 225, 250, 275
D	3	325, 350, 375
B	11	425, 450, 475, 500, 525, 550, 575, 600, 625, 650, 675
E	3	725, 750, 775
F	3	825, 850, 875
C	11	925, 950, 975, 1000, 1025, 1050, 1075, 1100, 1125, 1150, 1175
Total	42	

Table 4.1 The number of 1.23-MHz channels and the CDMA channel numbers whose center frequencies are preferred as the center frequencies of the 1.23-MHz frequency channels

First, note, in Table 4.1, that the preferred channels are separated by 1.25 MHz. A 1.23-MHz frequency channel is then created using the center frequency of one of these preferred channels as the center frequency of the 1.23-MHz channel. For example, consider Channel 25 and Channel 50 in the A-block. The center frequency of Channel 25 is 1851.25 MHz on the reverse link and 1931.25 MHz on the forward link; and the center frequency of Channel 50 is 1852.50 MHz on the reverse link and 1932.50 MHz on the forward link.

The 1.23-MHz frequency channels created from these two channels are shown in Figure 4.3 for the reverse and forward links. As shown in Table 4.1, a maximum of 42 1.23-MHz frequency channels may be created from the 60 MHz of bandwidth in each direction.

The Creation of the Code Channels

The final step in creating physical channels is to use the CDMA technique to create multiple channels from each of these 1.23-MHz-wide frequency channels. These final channels created by the CDMA technique are referred to as the code channels as opposed to the frequency channels. In this system, the code channels are created by the Direct Sequence (DS) spread spectrum technique.

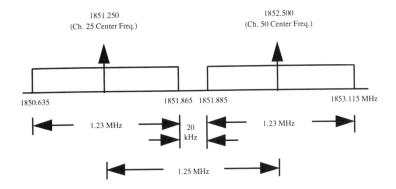

The Reverse Link

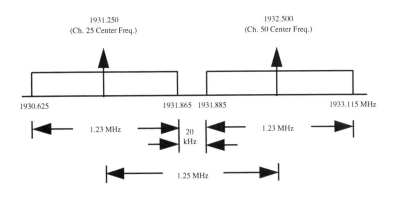

The Forward Link

Figure 4.3 The illustration of frequency channels created from Channel 25 and Channel 50

All of the individual code channels created for a frequency channel shares the entire 1.23-MHz bandwidth of that frequency channel, operating at the same carrier frequency, the center frequency of that frequency channel. Each code channel is uniquely identified by a code, or a sequence of "chips." The rate at which the coded or "spread" sequence of data called chips are transmitted is referred to as the chip rate. The chip rate in this system is 1.2288 Mcps. These 1.2288-Mcps code channels are the ultimate physical channels of this system. Figure 4.4 illustrates the concept of the code channel sharing the same frequency channel.

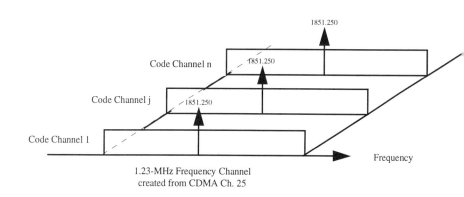

Figure 4.4   The illustration of the creation of code channels from a 1.23-MHz frequency channel

On the reverse link, the code channels are created by spreading the original source signal with a code referred to as the "long code" with a period $2^{42} -1$ at the chip rate of 1.2288 Mcps. On the forward link, the code channels are created by spreading the signal using the Walsh orthogonal codes. There are 64 different codes in the Walsh code at the chip rate of 1.2288 Mcps and, therefore, 64 physical channels of 1.2288 Mcps are created for the forward link.

4.3 LOGICAL CHANNELS

There are three logical channels on the forward link and two on the reverse link as follows:

- Forward link logical channels

  - *Synchronization channel.* The synchronization channel is used by the personal station to acquire the initial time synchronization.

  - *Paging channel.* The paging channel is used by the base station to transmit system overhead information and mobile station specific messages to the mobile station.

  - *Forward traffic channel.* The forward traffic channel is used by the base station for the transmission of user and signaling information to a specific mobile station during a call.

- Reverse link logical channels

  - *Access channel.* The access channel is used by the mobile station to initiate communication with the base station and to respond to paging channel messages.

  - *Reverse traffic channel.* The reverse traffic channel is used for the transmission of user and signaling information to the base station during a call.

The transmission rates of these logical channels are 1.2 kbps on the synchronization channel, fixed at either 9.6 or 4.8 kbps on the paging channel, and 4.8 kbps on the access channel. The transmission rate of the traffic channel is variable.

Two sets of transmission rates are defined for the traffic channel. Rate Set 1 is required and includes the following four rates: 9.6, 4.8, 2.3 and 1.2 kbps. Rate Set 2 is optional, and includes the following four rates: 14.4, 7.2, 3.6 and 1.8 kbps. The transmission rates are the same on the forward and reverse traffic channels. Because of the overhead bits included in the data frames as will be discussed later, the effective information transmission rate on the access channel and the traffic channels are less than the total channel transmission rates.

Table 4.2 summarizes the total transmission rates and the information transmission rates for the logical channels.

Logical Channels	Trans. Rates (kbps)	Logical Channels	Trans. Rates (kbps)
REVERSE LINK		**FORWARD LINK**	
Access Channel	4.8	**Synch Channel**	1.2
		Paging Channel	9.6
ReverseTraffic Channel		**Forward Traffic Channel**	
	Rate Set 1		Rate Set 1
	9.6		9.6
	4.8		4.8
	2.4		2.4
	1.2		1.2
	Rate Set 2		Rate Set 2
	14.4		14.4
	7.2		7.2
	3.6		3.6
	1.8		1.8

Table 4.2 The total transmission rates and the effective information transmission rates of the logical channels

4.4 FRAME FORMATS

A frame is a block of data bits transmitted in a fixed amount of time referred to as the frame duration. In this system, the frame duration is 20 *ms* for all logical channels except for the synchronization channel. For the synchronization channel, an 80-*ms* long superframe is defined, and the superframe is then divided into three frames. Therefore, the frame duration for the synchronization channel is 26.6.. *ms* or 80/3 *ms* to be exact. The number of data bits contained in a frame is referred to as the frame length and is equal to the transmission rate times the frame duration. For example, since the transmission rate of the access channel is 4.8 kbps and the frame duration is 20 *ms*, the access channel frame length is 96 bits.

Table 4.3 summarizes the frame durations and the frame lengths of the logical channels.

Logical Channels	Transmission Rates (kbps)	Frame Durations (ms)	Frame Lengths (bits)
REVERSE LINK			
Access Channel	4.8	20	96
Traffic Channel	Rate Set 1		
	9.6	20	192
	4.8	20	96
	2.4	20	48
	1.2	20	24
	Rate Set 2		
	14.4	20	288
	7.2	20	144
	3.6	20	72
	4.8	20	96
FORWARD LINK			
Synchronization Channel	1.2	26.66..	32
Paging Channel	9.6	20	192
	4.8	20	96
	1.8	20	36
Forward Traffic Channel	Rate Set 1		
	9.6	20	192
	4.8	20	96
	2.4	20	48
	1.2	20	24
	Rate Set 2		
	14.4	20	288
	7.2	20	144
	3.6	20	72
	4.8	20	96

Table 4.3 The frame durations and frame lengths of logical channel frames

The content of a frame is divided into fields. The lengths of the fields in a frame add up to the frame length. In the following subsections, the fields of the logical channel frames are discussed.

4.4.1 The Access Channel Frame

Figure 4.5 shows the access channel frame format. The access channel frame is 20 *ms* in duration, is 96 bits long, and contains the following two fields:

■ *Information (I) field*: The I-field contains the information bits transmitted by the personal station to the base station. The length of the I-field is 88 bits, and thus the effective information transmission rate on the access channel is 88 bits per frame duration, i.e., 20 *ms*, or 4.4 kbps.

■ *Encoder tail (T) field:* The length of the T-field is eight bits. These bits are appended to the I-field bits and are used by the convolutional encoder to reset itself to a known state.

Figure 4.5 The frame format of the access channel

4.4.2 The Traffic Channel Frames

The frame formats of the reverse and forward traffic channels are similar and are discussed together in this subsection. The frame duration is 20 *ms* all traffic channels and the frame length varies with the transmission rate of the traffic channel. Figure 4.6 shows the general format of the traffic channel frames. The specific frame format, however, depends on the transmission rate of the traffic channel. There are eight different transmission rates for the traffic channel: 9.6, 4.8, 2.4 and 1.2 kbps of

Rate Set 1 and 14.4, 7.2, 3.6 and 1.8 kbps of Rate Set 2. The I- and T-fields are present in all traffic channel frames. The E/R-field is present in all Rate Set 2 frames and is absent in all Rate Set 1 frames. The F-field is present in all traffic channel frames except the lowest two Rate Set 1 frames, 2.4 and 1.2 kbps frames.

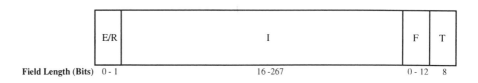

Figure 4.6 The general format of the traffic channel frames

The four fields of the traffic channel frame are discussed below:

- *Erasure indicator/reserved (E/R) field*: The length of the E/R-field is either one bit or zero, where zero-length field means that the field is absent. The E/R-field is present only in Rate Set 2 frames and is absent in Rate Set 1 frames. This field is called the E-field on the reverse traffic channel frame and is used to indicate an erased forward traffic channel frame to the base station. This same field is called the R-field on the forward traffic channel frame and is reserved for any future application.

- *Information (I) field*: The I-field is present in both the reverse and forward traffic channel frames of all eight transmission rates. The length of the I-field ranges from 16 bits for the lowest rate, 1.2 kbps, frame to 267 bits for the highest rate, 14.4 kbps, frame. The I-field carries the information payloads. Different types of information payloads may be multiplexed in the subfields of the I-field. The subfields of the I-field and multiplexing of information payloads will be discussed later.

- *Frame quality indicator (F) fled*: The F-field contains the Cyclic Redundancy Check (CRC) bits. The F-field is present in all Rate Set 2 frames and the highest two Rate Set 1 (i.e., 9.6 and 4.8 kbps) frames, and is absent in the lowest two

Rate Set 1 (2.4 and 1.2 kbps) frames. When present, the length of the F-field ranges from 6 bits for the 1.8 kbps frame to 12 bits for the 14.4 kbps frame.

- *Encoder tail (T) field*: The T-field is present in all traffic channel frames and is always eight-bits long.

Table 4.4 summarizes the field lengths and the frame lengths of the traffic channel frames of the eight different transmission rates. When a field is absent, its field length is shown as zero. The last column of the table shows the information bit transmission rate, i.e., the I-field length divided by the frame duration, 20 *ms*.

Trans. Rates (kbps)	E/R	I	F	T	Frame Length (bits)	I-Bit Trans. Rate (kbps)
		Field Length (bits)			*Frame Length (bits)*	*I-Bit Trans. Rate (kbps)*
Rate Set 1						
9.6	0	172	12	8	192	8.6
4.8	0	80	8	8	96	4
2.4	0	40	0	8	48	2
1.2	0	16	0	8	24	0.8
Rate Set 2						
14.4	1	267	12	8	288	13.35
7.2	1	125	10	8	144	6.25
3.6	1	55	8	8	72	2.75
1.8	1	21	6	8	36	1.05

Table 4.4 The field lengths and the frame lengths of the traffic channel frames

Information Multiplexing in the Traffic Channel I-Frames

Three different types of information may be multiplexed in the I-fields of the traffic channel:

- Primary traffic

■ Signaling traffic

■ Secondary traffic

The multiplex schemes for Rate Set 1 is referred to as the Multiplex Option 1; and that for Rate Set 2, the Multiplex Option 2. The main features of these two multiplex options are summarized below:

The Multiplex Option 1 for Rate Set 1

1. Under this multiplex option, the lower three Rate Set 1, i.e., 4.8, 2.4, 1.2 kbps, frames are not allowed to carry other than the primary traffic. For these frames, the I-field is not subdivided and contains 80, 40 and 16 bits of primary traffic (see Table 4.4). The I-fields of the lower three Rate Set 1 frames have no subfields and are same as those in Table 4.4.

2. With the highest Rate Set 1, i.e., 9.6 kbps, frame, one of the other two types of traffic can be multiplexed with the primary traffic or can be carried by itself without the primary traffic. Two types of multiplexing is specified for the 9.6 kbps frame: one is required and the other is optional. Under the required feature the frame may multiplex the primary traffic with the signaling traffic, or carry only the primary or the signaling traffic. Under the optional feature, the frame may multiplex the primary traffic with the secondary traffic, or carry only the secondary traffic.

3. With both the required and optional features of the Multiplex Option 1, only two types of traffic are multiplexed, i.e., the primary traffic with the signaling or the secondary traffic. As will be discussed later, it is possible to multiplex all three types of traffic with the Multiplex Option 2 for Rate Set 2.

4. In order to control the different multiplexing schemes, the first few bits of the I-field are used. The remainder of the I-field is divided to carry the different types of traffic. Figure 4.7 shows the general format of the I-field of the 9.6 kbps traffic channel. As shown in Figure 4.7 the first three subfields are used to control the multiplexing and the remaining subfields, to carry different types of traffic. As discussed below, when the MM bit is set to "0," the other two subfields, the TT- and TM-subfields, are absent in the I-field:

– *Mixed mode (MM) subfield*: The MM-subfield has length one bit, and is used to indicate whether the remainder of the I-field multiplexes the primary traffic with any other type of traffic. If the

MM bit is set to "0," the I-field carries only the primary traffic and the remainder of the I-field has no other subfields except the primary traffic subfield, i.e., the TT- and the TM-subfield discussed below are absent. If the MM bit is set to "1," the primary traffic is multiplexed with one of the other two types of traffic.

–   *Traffic Type (TT) subfield*:   The TT-subfield has length one bit, and is used to indicate whether the primary traffic is multiplexed with the signaling traffic (TT bit = "0"), or with the secondary traffic (TT bit = "1").

–   *Traffic mode (TM) subfield*:   The TM-subfield has length two bits, and is used to indicate four different mixes of the primary traffic with the other type of traffic. The amount of the primary traffic included in these four types of mixes are 80, 40, 16 and 0 bits per I-field of the frame. The first three numbers correspond to the numbers of the I-field bits of the lower three Rate Set 1, 4.8, 2.4, and 1.2 kbps, frames. The last mix, i.e., zero bit per I-field, corresponds to the case where no primary traffic is carried and only the signaling traffic (if TT bit = "0") or the secondary traffic (if TT bit = "1") is carried in the I-field.

–   Recall that the I-field length of the 9.6 kbps traffic channel frame is 172 bits. With the MM bit = "0," the remainder of the I-field is 171 bits. With the MM bit = "1," the I-field also has the TT- and TM-subfields for controlling the multiplexing scheme, amounting to a total of four bits, leaving 168 bits for the multiplexed traffic. With the MM bit = "1," the amount of the other type of traffic multiplexed with the primary traffic is, therefore, 88, 128, 152 and 168 bits, respectively, for the primary traffic of 80, 40, 16 and 0 bits per I-field of the frame.

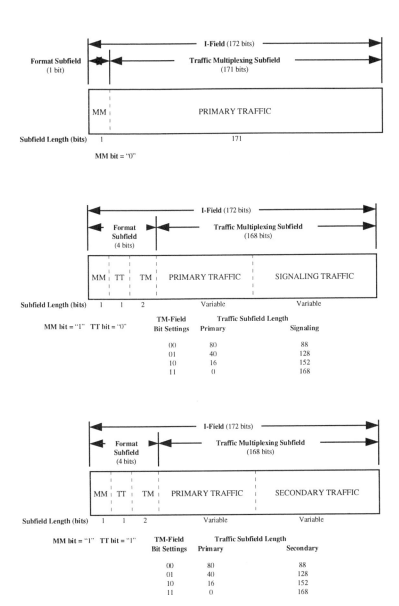

Figure 4.7 The I-field formats of the 9.6 kbps traffic channel frame for the Multiplex Option 1

The Multiplex Option 2 for Rate Set 2

1. Under the Multiplex Option 2 applied for the Rate Set 2 frames, the 14.4 and 7.2 kbps frames can multiplex up to all three types of traffic; the 3.6 kbps frame can multiplex up to two types of traffic - the primary traffic with the signaling traffic or with the secondary traffic; and, finally, the 1.8 kbps frame can carry only one type of traffic - either the primary traffic or the secondary traffic.

2. Two types of multiplexing is specified under this option: one required and the other optional. The choice of carrying only the primary traffic without multiplexing it with any other traffic is required for all Rate Set 2 frames. In addition, the 14.4, 7.2 and 3.4 kbps frames can carry the signaling traffic multiplexed with the primary traffic, or the signaling traffic singly without multiplexing it with the primary traffic. No such choice is specified for the 1.8 kbps frame: the only required feature for the 1.8 kbps frame is to carry the primary traffic only.

3. The optional multiplexing feature is similar to the required multiplexing feature described above except that, in this case, the secondary traffic instead of the signaling traffic is multiplexed with the primary traffic. Under this feature, the 14.4, 7.2 and 3.6 kbps frames can carry the secondary traffic multiplexed with the primary traffic or the secondary traffic only without multiplexing it with the primary traffic. The only optional feature for the 1.8 kbps frame is to carry the secondary traffic singly without multiplexing it with any other traffic.

4. One other feature of the optional multiplexing feature is available for the 14.4 and 7.2 kbps frames. These frames can multiplex all three types of traffic.

5. In order to control the different multiplexing schemes, the first few bits of the I-field are used. The remainder of the I-field is divided to carry the different types of traffic. Figures 4.8 through 4.11 show the general formats of the I-fields of the Rate Set 2 frames.

    –   *Mixed mode (MM) subfield*: The MM-subfield has length one bit, and is present in the I-fields of all Rate Set 2 frames. The MM-subfield is used to indicate whether the remainder of the I-field multiplexes the primary traffic with any other type of traffic. If the MM bit is set to "0," the frame carries only the primary traffic and the remainder of the I-field has no other subfields except the primary traffic subfield, i.e., the FM-subfield discussed below is absent. Figure 4.8 shows the general format of the I-fields of the Rate Set 2 frames when the MM bit is set to "0."

   –   *Frame mode (FM) subfield*: If the MM bit is set to "1," the primary traffic is multiplexed with the other types of traffic and the FM-subfield is present in the I-fields except in the I-field of the 1.8-kbps frame. Figures 4.9 through 4.11 show the general formats of the I-fields of the 14.4-, 7.2- and 3.6-kbps frames when the MM bit is set to "1." The FM-subfield is absent in the 1.8-kbps frame: the I-field of the 1.8-kbps frame with MM bit = "1" has the same format as that with MM bit ="0 shown in Figure 4.8 except that in this case the frame carries 20 bits of the secondary traffic instead of the primary traffic.

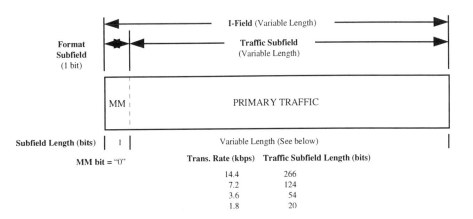

Figure 4.8 The I-field format of the Rate Set 2 frames for the Multiplex Option 2 with MM bit = "0"

   –   Referring to the figures, the length of the FM-subfield is four bits in the 14.4-kbps frame, three bits in the 7.2-kbps frame and two bits in the 3.6-kbps frame. The FM-subfield serves the combined function of the TT- and TM-subfields of the 9.6-kbps frame under the Multiplex Option 1 discussed earlier: it is used to indicate whether the primary traffic is multiplexed with the signaling traffic, with the secondary traffic, or with both the signaling and secondary traffic, and also is used to indicate different mixes of traffic when multiplexing. The various traffic mixes and the corresponding FM-field bit settings are shown in Figures 4.9 through 4.11.

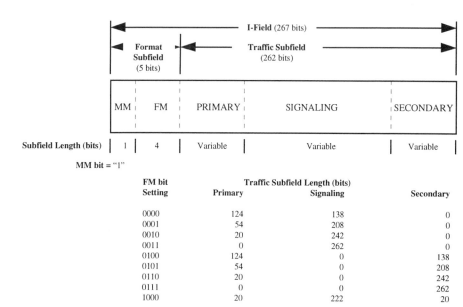

FM bit Setting	Traffic Subfield Length (bits)		
	Primary	Signaling	Secondary
0000	124	138	0
0001	54	208	0
0010	20	242	0
0011	0	262	0
0100	124	0	138
0101	54	0	208
0110	20	0	242
0111	0	0	262
1000	20	222	20

Figure 4.9 The I-field format of the 14.4-kbps frame for the Multiplex Option 2 with MM bit = "1"

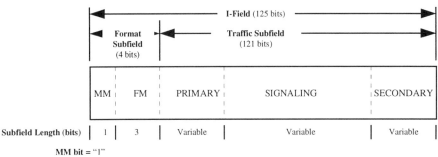

Figure 4.10 The I-field format of the 7.2-kbps frame for the Multiplex Option 2 with MM bit = "1"

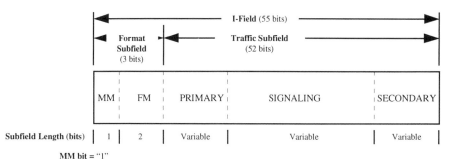

Figure 4.11 The I-field format of the 3.6-kbps frame for the Multiplex Option 2 with MM bit = "1"

4.4.3 The Synchronization Channel Frame

The synchronization channel frames are transmitted in groups of three frames called superframes. The superframe duration is 80 *ms*, and the frame duration is 80/3 *ms* or 26.66..*ms*. The transmission rate of the synchronization channel is 1.2 kbps and the lengths of the frame and the superframe are 32 bits and 96 bits, respectively. Figure 4.12 shows the format of the synchronization channel frame. The synchronization channel frame has two fields: the Start of Message (SOM) field and the frame body field. The SOM-field length is one bit; and the frame body field length, 31 bits. The synchronization channel carries only one type of message, the synchronization channel message. This message is broken into pieces and carried in the synchronization channel frames. The frame which carries the beginning of the message has the SOM-field bit set to 1. The rest of the frames have the SOM-field bit set to 0. The frame body field carries the message pieces.

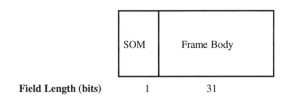

Figure 4.12   The frame format of the synchronization channel

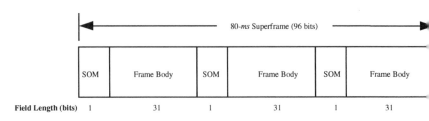

Figure 4.13   The superframe of the synchronization channel

4.4.4 The Paging Channel Frame

The paging channel transmission rate is fixed at 9.6 or 4.8 kbps. In other words, in a given system, the transmission rate is the same for all paging channels. The frame duration for the paging channel is 20 *ms*. Hence, the frame length is 192 bits for the 9.6-kbps paging channel, and 96 bits for the 4.8-kbps paging channel.

The paging channel is divided into paging channel slots 80 ms long each. Each paging channel slot contains four paging channel frames. Each frame is divided into two half frames. According to our convention of terminology, each half frame might be considered as a field of the paging channel frame. Each half frame is then divided into two subfields: the Synchronized capsule Indicator (SCI) subfield and the half frame body subfield.

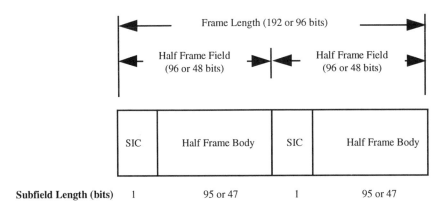

Figure 4.14 The frame format of the paging channel

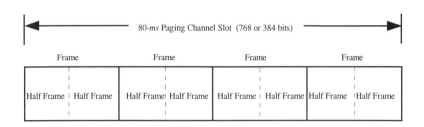

Figure 4.15   The paging channel slot

4.5   DIGITAL INFORMATION PROCESSING

The speech signal is sampled, quantized and represented by a series of "0's" and "1's" or bits. This bit stream is the original information source. In addition, the original information source may be digital data which are already in the form of bits. These information bits, speech or data, are put into the frames of the logical channels that carry the information according to the formats discussed in the previous sections. These frames are channel-coded, repeated appropriately number of times, interleaved and spread by codes. Finally, these coded information, or "chips," which are spread to fill an entire 1.23-MHz frequency channel are used to modulate the microwave carrier frequency, i.e., the center frequency of the 1.23-MHz frequency channel. This series of digital signal processing steps - frame formation, channel coding, symbol repetition, interleaving and code spreading - is depicted in Figure 4.16 and is discussed in the next several subsections.

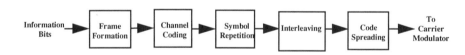

Figure 4.16   The digital signal processing

4.5.1 Frame Formation

The bits representing the speech or data signal are the information bits and are formatted into the appropriate logical channel frames according to the frame formats discussed earlier. The information bits are put into the I-field, and the various bits making up the other fields of the frame, such as the E/R-, F- and T-fields, are added to complete the frame. Table 4.4 shows the numbers of bits in the various fields of a frame. In addition, if the information bits on a traffic channel are a multiplexed traffic, the I-field should also include subfields such as the MM-, TT-, TM and FM-subfields, depending on the transmission rate. The numbers of bits added in these subfields of the I-field are shown in Figures 4.7 through 4.11.

For the synchronization channel and the paging channel, the frame contains only the information bits, i.e., only the I-field, and does not contain any other field. The final outputs of this frame formation process are bit streams of the various transmission rates: Rate Set 1 (9.6, 4.8, 2.4, and 1.2 kbps) and Rate Set 2 (14.4, 7.2, 3.6, and 1.8 kbps) for the traffic channel (reverse and forward), 4.8 kbps for the access channel, 1.2 kbps for the synchronization channel, and either 9.6 or 4.8 kbps for the paging channel.

4.5.2 Channel Coding and Symbol Repetition

The input to the channel coding process is the formatted bit stream, i.e., the frames, of the logical channel. For all of the logical channels, the data frames are convolutionally encoded with a constraint length $K = 9$. Two different code rates are used: $r = 1/2$ or $1/3$. After the encoding, the symbols are repeated appropriate numbers of times to yield common symbol rates for the logical channels.

Table 4.5 summarizes the code rates used for the various logical channels, the symbol transmission rates at the output of the encoder, the numbers of times the encoded symbols are repeated, and the final symbol transmission rate after the symbol repetition. On the reverse link, the symbol transmission rate after the repetition is 28.8 ksps for all of the logical channels. On the forward link, two different symbol transmission rates result after the symbol repetition (and, in addition, puncturing for Rate Set 2 frames): for the synchronization channel, it is 4.8 ksps; and for the rest of the logical channels, it is 19.2 ksps. These are the symbol transmission rates at the input to the interleaver.

Logical Channels	Trans. Rates @ Coder Input (kbps)	Code Rate, r	Sym. Rates @ Coder Output (ksps)	No. of Code Repe- tition	Sym. Rates after Repe- tition (ksps)
REVERSE LINK					
Access Channel	4.8	1/3	14.4	2	28.8
Reverse Traffic					
Channel	Rate Set 1				
	9.6	1/3	28.8	1	28.8
	4.8	1/3	14.4	2	28.8
	2.4	1/3	7.2	4	28.8
	1.2	1/3	3.6	8	28.8
	Rate Set 2				
	14.4	1/2	28.8	1	28.8
	7.2	1/2	14.4	2	28.8
	3.6	1/2	7.2	4	28.8
	1.8	1/2	3.6	8	28.8
FORWARD LINK					
Synch. Channel	1.2	1/2	2.4	2	4.8
Paging Channel	9.6	1/2	19.2	1	19.2
	4.8	1/2	9.6	2	19.2
Forward Traffic					
Channel	Rate Set 1				
	9.6	1/2	19.2	1	19.2
	4.8	1/2	9.6	2	19.2
	2.4	1/2	4.8	4	19.2
	1.2	1/2	2.4	8	19.2
	Rate Set 2				
	14.4	1/2	28.8	1	28.8 (19.2)*
	7.2	1/2	14.4	2	28.8 (19.2)*
	3.6	1/2	7.2	4	28.8 (19.2)*
	1.8	1/2	3.6	8	28.8 (19.2)*

* The number if the parentheses is the symbol rate after puncturing.

Table 4.5   The frame durations and frame lengths of logical channel frames

4.5.3 Interleaving

After the repetition (and, in the case of Rate Set 2 frames for the forward traffic channel, after the repetition and subsequent puncturing), the symbols are interleaved by the block interleaver. Interleaving is performed on a block of symbols transmitted over one frame duration, i.e., 20 *ms* for all of the reverse and forward logical channels except for the synchronization channel for which the frame duration is 26.66...*ms* (80/3 *ms* to be exact).

For all reverse link logical channels, the symbol rate at the input to the interleaver is 28.8 ksps and, therefore, there are 576 symbols to interleave for the duration of 20 *ms*; for all forward link logical channels except for the synchronization channel, the symbol rate at the input to the interleaver is 19.2 ksps, and there are 384 symbols to interleave for the duration of 20 *ms*; and, finally, for the synchronization channel, the symbol rate at the input to the interleaver is 4.8 ksps, and there are 128 symbols to interleave for the duration of 26.66..ms; and, finally,

The interleaving is accomplished by the "write" and "read" operations. The block of symbols of one frame duration are first written into an *nxm* matrix and then read out from the matrix in a different order. The dimensions of the matrix, *n* and *m,* are specified for the three different symbol transmission rates discussed above, such that the number of entries, or cells, in the matrix, i.e., *n* times *m*, is equal to the number of symbols to be interleaved. The dimensions of the matrix for all of the reverse link logical channels are 32x18 with 576 cells. On the forward link, the dimensions are 24x16 with 384 cells for all forward traffic channels and paging channels and 16x8 with 128 cells for the synchronization channel.

The write operation is same on both the reverse and forward links: the symbols are written into the matrix column by column. The read operation is performed as follows. For the reverse link, the interleaver keeps one matrix. The symbols are read out from the same matrix row by row. For the forward link, two matrices are kept: the "write matrix" is same as that for the reverse link. The "read matrix" is created by rearranging the elements of the write matrix according to a certain algorithm. The read operation is discussed below for the reverse link first and then for the forward link.

The Read Operation for
the Reverse Link Channels

The order of rows by which the symbols are read out for the reverse traffic channel is determined by the following algorithm depending on the transmission rate. For the transmission rate of 9.6 and 14.4 kbps, the order is simply the natural order of the row addresses from the first row to the thirty-second row. For the transmission rate of 4.8 and 7.2 kbps, the order is determined as follows. First, the 32 rows are divided into eight groups with four rows per group, each of which is then divided into two subgroups with two rows per subgroup. Starting from the first group of four rows, the rows are read out in the natural order in each subgroup alternating between the two subgroups. For example, with the first group of four rows, the readout order is: the first row of the first subgroup, i.e., row 1, the first row of the second subgroup, i.e., row 3, the second row of the first subgroup, i.e., row 2 and, finally, the second row of the second subgroup, i.e., row 4.

Once one group of four rows is read out completely in this manner, the process is repeated with the next group of four rows. Similarly, for the transmission rate of 2.4 and 3.6 kbps, the 32 rows are divided into four groups with eight rows per group, each of which is then divided into two subgroups with four rows per subgroup. Finally, for the transmission rate of 1.2 and 1.8 kbps, the 32 rows are divided into two groups with 16 rows per group, each of which is divided into two subgroups with eight rows per subgroup. The same basic algorithm is then applied to these groups and subgroups to determine the readout order. For the access channel, the algorithm is the bit-reversed readout of the row addresses. First, the 32 row addresses from 0 to 31 are expressed in the binary number, i.e., 00000 for 0, 00001 for 1,..., 11111 for 31. These bits are reversed. Let n be the bit-reversed binary number. Then the row address is n+1.

The final readout orders of the rows determined in this manner for the reverse link logical channels are shown in Table 4.6.

Logical Channels	*Row Order in Data Readout*
Rev. Traff Chan. (kbps)	
9.6/14.4	1, 2, 3, 4, 5, 6, 7, 8, 9, 10, 11, 12, 13, 14, 15, 16, 17, 18, 19, 20, 21, 22, 23, 24, 25, 26, 27, 28, 29, 30, 31, 32
4.8/7.2	1, 3, 2, 4, 5, 7, 6, 8, 9, 11, 10, 12, 13, 15, 14, 16, 17, 19, 18, 20, 21, 23, 22, 24, 25, 27, 26, 28, 29, 31, 30, 32
2.4/3.6	1, 5, 2, 6, 3, 7, 4, 8, 9, 13, 10, 14, 11, 15, 12, 16, 17, 21, 18, 22, 19, 23, 20, 24, 25, 29, 26, 30, 27, 31, 28, 32
1.2/1.8	1, 9, 2, 10, 3, 11, 4, 12, 5, 13, 6, 14, 7, 15, 8, 16, 17, 25, 18, 26, 19, 27, 20, 28, 21, 29, 22, 30, 23, 31, 24, 32
Access Channel	1, 17, 9, 25, 5, 21, 13, 29, 3, 19, 11, 27, 7, 23, 15, 31, 2, 18, 10, 26, 6, 22, 14, 30, 4, 20, 12, 28, 8, 24, 16, 32

Table 4.6   The row order of data readout for the reverse traffic channel

The Read Operation for the Forward Link Channels

The read matrix is created by filling the elements, or cells, of the matrix column by column with the elements of the write matrix according to certain algorithms. Once the read matrix is created in this manner, the cells of the read matrix are read out column by column in the natural order of the columns. The algorithms of filling the read matrix columns from the write matrix are discussed below for the synchronization channel, the forward traffic channels and the paging channels.

The dimensions of the write matrix for the synchronization channel are 16x8. The read matrix for the synchronization channel has the same dimensions, 16x8. The columns of this matrix are filled by the elements (i, j) of the write matrix according to the following algorithm. First, the row order i is determined by grouping, subgrouping, and sub-subgrouping the 16 rows, and alternating between them to be as follows: 1, 9, 5, 13, 3, 11, 7, 15, 2, 10, 6, 14, 4, 12, 8, 16. Once a row i is selected according to the above order, all of the elements of that row are read into the read matrix according to the order of the column index j which is determined as follows. For a given row, the eight elements of the row are divided into two groups with four elements per group, each of which is divided into two subgroups of two elements each and the column index j is determined according to a similar algorithm alternating between the groups to be as follows: 1, 5, 3, 7, 2, 6, 4, 8. To determine the order of elements of the write matrix (i, j) to fill the read matrix with, for a given i, apply the above order of j, for all i in the order i determined above. For example, starting with $i = 1$, (1, 1), (1, 5), (1, 3), (1, 7), (1, 2), (1, 6), (1, 4), (1, 8) of the write matrix are used to fill the first eight cells of the first column of the read matrix. The next eight cells are filled with the next i in order $i = 9$ and the above order of j, i.e., (9, 1), (9, 5), ... , (9, 8).

For the forward traffic channels and the paging channels, the dimensions of the write and read matrices are both 24x16 with 384 elements, or cells. To form the read matrix, first put these 384 elements in series by reading out from the write matrix column by column starting with the first column. These 384 elements are divided into groups, subgroups, etc. similarly. The order is determined element by element (i, j), and the elements of the write matrix in this specific order are then used to fill the read matrix column by column starting with the first column.

4.5.4 Spreading

The Spreading for
the Reverse Link Channels

On the reverse link, the output of the block interleaver is 28.8 ksps for both the access channel and all of the reverse traffic channels. These 28.8 ksps streams undergo three steps of code spreading as follows. The same spreading technique is applied for the access channel and the reverse traffic channels. The first step is to spread the 28.8 ksps stream by the Walsh function. The Walsh code has 64 distinct codes. Therefore, each Walsh code can represent six symbols, i.e., $2^6 = 64$. The stream of symbols at 28.8 ksps is divided into groups of six and each group of six symbols is spread by 64 chips of a Walsh code. Hence, the output of this process is a stream of coded symbols at the chip rate of 307.20 kcps. This process is referred to as the 64-ary orthogonal modulation because the Walsh codes are orthogonal and the 28.8 ksps symbol stream is used to modulate the Walsh function.

The next step is to spread the 307.2 Mcps stream with the "long codes," which has a period of $2^{42} - 1$ chips. Each Walsh chip at the rate 307.2 Mcps is coded by four chips of the long code. After this spreading by a long code, the chip rate for the logical channels is 1.2288 Mcps.

The final step of the spreading is the quadrature spreading. This system uses the Quadrature Phase Shift Keying (QPSK) modulation. in the QPSK, four different phases are used in the modulation and each phase of the carrier can represent two symbols, or in this case two chips. Ordinarily, therefore, in a non-spread spectrum system, the symbol rate would be reduced to one half before the symbol stream is used to modulate the phase of the carrier. In a spread spectrum system, the symbol rate is not reduced, which is again equivalent to "spreading."

In this system, the stream of chips at 1.2288 Mcps are duplicated into two parallel branches, the I- and Q-channels, and are coded by different long codes, the I- and Q-sequences, at the same chip rate, 1.2288 Mcps. The result are two parallel streams of 1.2288 Mcps. A pair of chips, one from the I-channel and the other from the Q-channel, are then used to modulate the phase of the carrier. The mapping between the (I, Q) symbol pair and the phase angle of the modulated carrier is show below:

I	Q	Phase Angle
0	0	$\pi/4$
1	0	$3\pi/4$
1	1	$-3\pi/4$
0	1	$-\pi/4$

The Spreading for the Forward Link Channels

The spreading on the forward link is similar to that discussed above for the reverse link. The forward link channels are spread by the Walsh function at the chip rate of 1.2288 Mcps. Therefore, there are a total of 64 forward link logical channels. The mapping between the (I, Q) symbol pair and the phase angle of the modulated carrier is show below.

I	Q	Phase Angle
0	0	$\pi/4$
1	0	$3\pi/4$
1	1	$-3\pi/4$
0	1	$-\pi/4$

4.6   CALL PROCESSING

4.6.1 The Personal Station Call Processing

There are four main states in which the personal station performs the call processing procedures: the personal station initialization state, the personal station idle state, the system access state, and the traffic channel control state.

The Call Processing in
the Personal Station Initialization State

The personal station moves into its initialization state when the user powers up the personal station. It may also move into this state from the personal station idle state at the end of an idle hand-off or from the traffic channel control state when the use of the traffic channel ends. While in the personal station initialization state, the personal station performs the following major processing functions:

1.   Initializes the registration parameters.

2.   Selects the system to use.

3.   Acquires the pilot channel of the selected CDMA system by setting its code channel for the pilot channel, tuning to the appropriate CDMA channel number, and searching for the pilot channel.

4. After acquiring the pilot channel, sets its code channel for the synchronization channel and receives *the Synchronization Channel Message.*

5. Processes the *Synchronization Channel Message* to obtain system configuration and timing information, and stores the information received in the *Synchronization Channel Message* including the system identifier (ID), the network ID, the pilot pseudo noise (PN) sequence offset, the long code state, the system time, and the paging channel data rate.

6. Finally, synchronizes its long code timing and system timing to those of the CDMA system using the information received from the *Synchronization Channel Message,* namely, the pilot PN sequence offset, the long code state, and the system time.

The Call Processing in the Personal Station Idle State

The personal station moves into the idle state from the initialization state when it successfully completes the initialization procedures. It may also move into this state from the system access state when it receives an acknowledgment after transmitting a message on the access channel while in the system access state unless the message it had sent is an *Origination Message* or a *Page Response Message.*

While in its idle state, the personal station performs the following major procedures:

1. *Paging channel monitoring procedures.* When the personal station moves into the idle state, it sets its code channel to the paging channel, sets the paging channel data rate to the value received in the *Synchronization Channel Message,* and starts monitoring the paging channel. The paging channel is divided into paging channel slots each 80-*ms* long containing eight paging channel half frames, or four paging channel frames.. The paging channel slots repeat in cycles, called slot cycles. The minimum length slot cycle consists of 16 slots and thus has a slot cycle period of 1.28 seconds. This slot cycle length may increase in multiples of 2, i.e., the next to the minimum slot cycle would be 2.56 seconds. A personal station may monitor the paging channel in a slotted mode or non-slotted mode. In the non-slotted mode, the personal station must monitor all slots; in the slotted mode, only the specified slots. In the slotted mode, the personal station typically monitors one or two slots per slot cycle, and therefore, the personal station can conserve power while it is not monitoring its slots.

2. *Acknowledgment procedures.* The personal station sends acknowledgments of messages received on the paging channel on the access channel.

3. *Registration procedures.* The personal station performs the registration procedures. More will be discussed on the registration procedures later.

4. *Idle hand-off procedures.* The idle hand-off refers to the hand-off of the personal station from one pilot channel to another while it is in an idle state, i.e., while it is not engaged in a call. An idle hand-off occurs if the personal station changes the coverage area of the base station while it is in its idle state. The personal station searches the pilot channels according to three different sets of pilots channels: the active set, the neighbor set, and the remaining set. The personal station performs an idle hand-off in the non-slotted mode.

5. *Response to overhead information operation.* The personal station performs this operation whenever it receives an overhead message on the paging channel and, if necessary, updates its memory according to the received message. Among the overhead messages are the *System Parameters Message*, the *Access Parameters Message*, and the *CDMA Channel List Message*. When the contents of one or more of these messages change, the configuration message sequence number contained in these messages is incremented. The personal station compares the value of this parameter as received in the message with the parameter's value as currently stored in the personal station. If the two match, the personal station ignores the overhead messages ending the operation; and, if they do not match, updates its memory.

6. *Personal station page match operation.* This is the procedure whereby the personal station determines whether there is any incoming call directed to it. Whenever the personal station receives a *General Page Message*, it searches the message to determine whether it contains the IMSI or TMSI assigned to itself. If it does, the personal station sends a *Page Response Message* to the base station on the access channel.

7. *Personal station order and message processing operation.* The paging channel is used to transmit messages intended not only for paging but also for other purposes. These other messages may be intended to order the personal stations to perform certain actions. The personal station performs this operation whenever it receives a message or order other than a *General Page Message*.

8. *Personal station origination operation.* The personal station performs this operation when it is directed by the user to initiate a call.

Personal station message transmission operation. If the personal station supports the transmission of *Data Burst Messages*, it performs this operation whenever the user directs it to transmit such a message.

). *Personal station power-down operation.* The personal station performs this operation when it is directed by the user to power down.

he Call Processing in
1e System Access State

he personal station moves to the system access state from its idle state when it eds to acquire the access channel to respond to a paging channel message, to riginate a call or to perform registration.

he access procedures

he messages sent on the access channel are of two types: a response to a base ation message, and a request by the personal station. The access attempt begins ith the personal station sending a message on the access channel to the base station. may need to send the same message many times until it receives an knowledgment. The access attempt ends when the personal station receives an knowledgment to the message from the base station. This entire process of nding one message and receiving (or eventually failing to receive for whatever ason) an acknowledgment is considered one access attempt. One access attempt ay consist of transmitting the same message multiple times and each transmission f the message is called an access probe. Each access probe consists of an access nannel preamble and an access channel message capsule.

ithin an access attempt, access probes are divided into groups of up to a maximum f 16 access probes. Each of these access probe groups is called an access probe quence. Therefore, one access attempt may transmit multiple access probe quences. Each access probe sequence is transmitted on a different access channel. he access channel on which a particular access probe sequence is transmitted is seudo-randomly selected from among all of the access channels associated with the urrent paging channel.

he access channel is divided into access channel slots. The transmission of an cess probe begins at the start of an access channel slot. More on the access nannel structure will be discussed later as part of the discussion on the signaling ɔrmats. The personal station transmits on the access channel using a random access procedure based on the parameters supplied by the base station in the *Access* *Parameters Message*. The timing of the start of each access probe sequence and the timing between access probes within an access probe sequence are both determined pseudo-randomly. After transmitting an access probe sequence, the next sequence is transmitted after a backoff delay, RS, which is determined pseudo-randomly. For the request message only, the adjacent access probe sequences are separated by an additional delay, PD, plus the backoff delay, RS. The precise timing of the access channel transmissions in an access attempt is determined as follows. The personal station computes a number, RN, which depends on its Electronic Serial Number (ESN) and delays its transmit timing by RN PN chips.

While in the system access state, the personal station monitors the paging channel until it has a current set of overhead messages and also sends a response to a message received from the base station. In addition, while in this state, the personal station may send the following specific messages to the base station:

- *Origination Message*
- *Page Response Message*
- *Registration Message*
- *Data Burst Message*

The Call Processing in
the Traffic Channel Control State

When directed to a traffic channel, the personal station moves into the traffic channel control state from the system access channel. In this state the personal station and the base station communicate using the reverse and forward traffic channels. While in this state, the personal station verifies that it can receive the forward traffic channel and begins transmitting on the reverse traffic channel, exchanges traffic channel frames with the base station for the conversation, and disconnects the call.

4.6.2 The Base Station Call Processing

The base station call processing consists of the following types of processing:

1. The base station transmits signals on the pilot channel and message frames on the synchronization channel so that the personal station may synchronize itself to the CDMA system while in the initialization state.

2. The base station transmits message frames on the paging channel which the personal station monitors to receive any incoming call while in the idle state.

3. The base station monitors the access channel to receive messages which the personal station sends while in the system access state.

4. The base station uses the forward and reverse traffic channels to communicate with the personal station while in the traffic channel control state.

4.6.3 Registration

This system supports the following nine different forms of registration:

1. Power-up registration

2. Power-down registration

3. Timer-based registration

4. Distance-based registration

5. Zone-based registration

6. Parameter-change registration

7. Ordered registration

8. Implicit registration

9. Traffic channel registration

4.7 SIGNALING MESSAGE FORMATS

4.7.1 Signaling on the Reverse Link

This subsection discusses the messages transmitted on the reverse logical channels, i.e., the access channel and the reverse traffic channel. The following is a complete list of the messages transmitted on the reverse link logical channels. They are listed in three groups: those transmitted on the access channel only; those, on the reverse traffic channel only; and those, on both the access channel and the reverse traffic channel:

The messages transmitted on the access channel only are:

- Registration Message
- Origination Message
- Page Response Message

The messages transmitted on the reverse traffic channel only:

- Flash with Information Message
- Pilot Strength Measurement Message
- Power Measurement Report Message
- Send Burst DTMF Message
- Reserved for Obsolete Status Message
- Origination Continuation Message
- Hand-off Completion Message
- Parameters Response Message
- Service Request Message
- Service Response Message
- Service Connect Completion Message
- Service Option Control Message

Messages transmitted on both the access and reverse traffic channels

- Order Message
- Data Burst Message
- Authentication Challenge Response Message
- Status Response Message
- TMSI Assignment Completion Message

Signaling on the Access Channel

The construction of an access probe

As discussed earlier, an access attempt generally consists of transmitting the same message multiple times until the attempt succeeds with receiving an acknowledgment or eventually fails. Each transmission containing the same message is called an access probe. The content of each probe is constructed as follows. An access channel message to be transmitted is put into an access channel message capsule. If the message does not fill the capsule length completely, it is appended with a padding. This access channel message capsule is then subjected to the frame formation process discussed earlier. The message capsule is broken into an integer number, say N, of segments and each segment is appended with the T-field bits. The result is a series of N access channel frames, each 96 bits in length and 20 ms in duration. Since the T-field length is 8 bits and the access channel frame length is 96 bits, the access channel message capsule length must be 88 bits times the number of frames, N. These frames are appended to an access channel preamble. An access channel preamble is a series of an integer number, say M, of access channel frames each consisting of 96 zeros. These M+N access channel frames constitute an access probe. Each probe is transmitted during one access channel slot. Figure 4.17 shows the relationships between an access channel message, an access channel message capsule, an access channel preamble, access channel frames and an access channel slot.

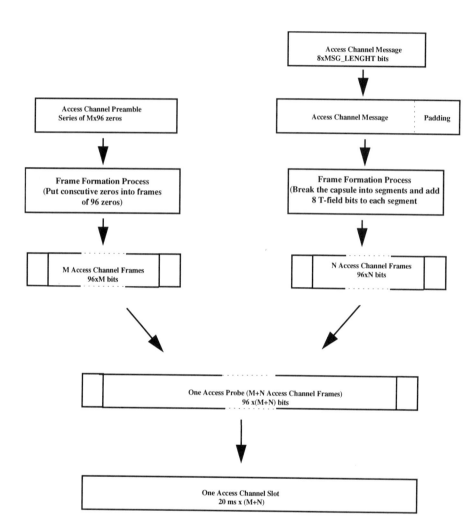

Figure 4.17 The construction of a access probe from an access channel preamble and an access channel message capsule

The format of the access channel message

The main element of an access probe is an access channel message, which is encapsulated in an access channel message capsule as discussed above. Figure 4.18

shows the format of an access channel message. Each message contains the following three fields:

■ *Message length field*: This field is eight bits long, and indicates the length in octets of the access channel message, i.e., the sum of the lengths of the three fields. The maximum length of an access channel message is 880 bits or 110 octets.

■ *Message body field*: This field contains the main body of the message. The length of this field is variable. Since the message length field is eight bits long and the CRC field is 30 bits long, the minimum length of the message body field is two bits yielding the message length of 40 bits, which is a multiple of eight bits, i.e., five octets (i.e., 38 bits of the message length field and the CRC field is not an integer multiple of eight bits). Since the maximum message length is 880 bits, the maximum length of the message body field is 842 bits.

■ *CRC field*: This field contains a 30-bit CRC computed for the access channel message. The CRC is computed for the message length and message body fields.

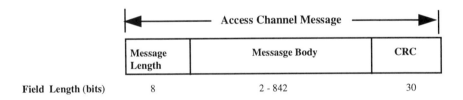

Figure 4.18 The format of an access channel message

The format of the message body field of the access channel message

The format of the message body field varies with the access channel message. There are eight different signaling messages transmitted on the access channel, i.e., the three listed among the first group and the five listed among the third group above. Figures 4.19 through 4.26 show the formats of the message body field of these eight messages.

Subfield Name	Subfield Length (bits)
Message Type (MSG_TYPE)	8
Acknowledgement Sequence Number (ACK_SEQ)	3
Message Sequence Number (MSG_SEQ)	3
Acknowledgement Required Indicator (ACK_REQ)	1
Valid Acknowledgement Indicator (VALID_ACK)	1
Acknowledgement Address Type (ACK_TYPE)	3
Personal Station Identifier Field Type (MSID _TYPE)	3
Personal Station Identifier Field Length (MSID_LEN)	4
Personal Station Identifier (MSID)	8xMSID_LEN
Authentication Mode (AUTH_MODE)	2
Authentication Data (AUTHR)	0 or 18
Random Challenge Value (RANDC)	0 or 8
Call History Parameter (COUNT)	0 or 6
Registration Type (REG_TYPE)	4
Slot Cycle Index (SLOT_CYCLE_INDEX)	3
Protocol Revision of the Personal Station (MOB_P_REV)	8
Extended Station Class Mark (EXT_SCM)	1
Reserved Bits (RESERVED)	1
Slotted Mode (SLOTTED_MODE)	1
Reserved Bits (RESERVED)	5
Pers. Station Terminated Calls Accepted Indicator (MOB_TERM)	1
Reserved Bits (RESERVED)	6

Message Body Field

Figure 4.19   The format of the message body field of the *Registration Message*

Subfield Name	Subfield Length (bits)
Message Type (MSG_TYPE)	8
Acknowledgement Sequence Number (ACK_SEQ)	3
Message Sequence Number (MSG_SEQ)	3
Acknowledgement Required Indicator (ACK_REQ)	1
Valid Acknowledgement Indicator (VALID_ACK)	1
Acknowledgement Address Type (ACK_TYPE)	3
Personal Station Identifier Field Type (MSID _TYPE)	3
Personal Station Identifier Field Length (MSID_LEN)	4
Personal Station Identifier (MSID)	8xMSID_LEN
Reserved Bits (RESERVED)	2
Order Code (ORDER)	6
Additional Record Length (ADD_RECORD_LEN)	3
Oder-specific subfields (if used)	8x ADD_RECORD_LEN
Reserved Bits (RESERVED)	5

Figure 4.20 The format of the message body field of the *Order Message*

	Subfield Name	Subfield Length (bits)
	Message Type (MSG_TYPE)	8
	Acknowledgement Sequence Number (ACK_SEQ)	3
	Message Sequence Number (MSG_SEQ)	3
	Acknowledgement Required Indicator (ACK_REQ)	1
	Valid Acknowledgement Indicator (VALID_ACK)	1
	Acknowledgement Address Type (ACK_TYPE)	3
	Personal Station Identifier Field Type (MSID _TYPE)	3
	Personal Station Identifier Field Length (MSID_LEN)	4
	Personal Station Identifier (MSID)	8xMSID_LEN
Message Body Field	Authentication Mode (AUTH_MODE)	2
	Authentication Data (AUTHR)	0 or 18
	Random Challenge Value (RANDC)	0 or 8
	Call History Parameter (COUNT)	0 or 6
	Message Number within Data Burst Stream (MSG_NUMBER)	8
	Data Burst Type (BURST_TYPE)	6
	Number of Messages in Data Burst Stream (NUM_MSGS)	8
	Number of Characters in Current Message (NUM_FIELDS)	8
	1st Character (CHAR-1)	8
	2nd Character (CHAR -2)	8
	.	.
	.	.
	NUM_FIELDSth Character (CHAR-NUMFIELDS)	8

Figure 4.21   The format of the message body field of the *Data Burst Message*

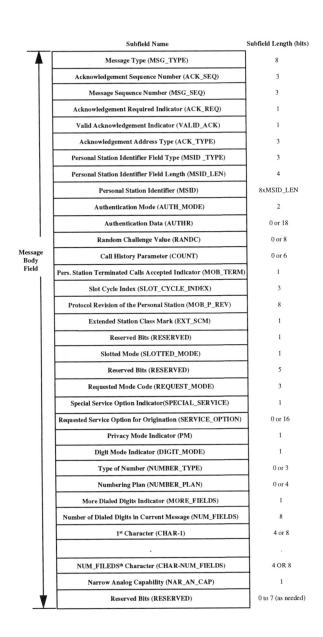

Subfield Name	Subfield Length (bits)
Message Type (MSG_TYPE)	8
Acknowledgement Sequence Number (ACK_SEQ)	3
Message Sequence Number (MSG_SEQ)	3
Acknowledgement Required Indicator (ACK_REQ)	1
Valid Acknowledgement Indicator (VALID_ACK)	1
Acknowledgement Address Type (ACK_TYPE)	3
Personal Station Identifier Field Type (MSID _TYPE)	3
Personal Station Identifier Field Length (MSID_LEN)	4
Personal Station Identifier (MSID)	8xMSID_LEN
Authentication Mode (AUTH_MODE)	2
Authentication Data (AUTHR)	0 or 18
Random Challenge Value (RANDC)	0 or 8
Call History Parameter (COUNT)	0 or 6
Pers. Station Terminated Calls Accepted Indicator (MOB_TERM)	1
Slot Cycle Index (SLOT_CYCLE_INDEX)	3
Protocol Revision of the Personal Station (MOB_P_REV)	8
Extended Station Class Mark (EXT_SCM)	1
Reserved Bits (RESERVED)	1
Slotted Mode (SLOTTED_MODE)	1
Reserved Bits (RESERVED)	5
Requested Mode Code (REQUEST_MODE)	3
Special Service Option Indicator(SPECIAL_SERVICE)	1
Requested Service Option for Origination (SERVICE_OPTION)	0 or 16
Privacy Mode Indicator (PM)	1
Digit Mode Indicator (DIGIT_MODE)	1
Type of Number (NUMBER_TYPE)	0 or 3
Numbering Plan (NUMBER_PLAN)	0 or 4
More Dialed Digits Indicator (MORE_FIELDS)	1
Number of Dialed Digits in Current Message (NUM_FIELDS)	8
1st Character (CHAR-1)	4 or 8
.	.
NUM_FILEDSth Character (CHAR-NUM_FIELDS)	4 OR 8
Narrow Analog Capability (NAR_AN_CAP)	1
Reserved Bits (RESERVED)	0 to 7 (as needed)

Message Body Field

Figure 4.22 The format of the message body field of the *Origination Message*

	Subfield Name	Subfield Length (bits)
	Message Type (MSG_TYPE)	8
	Acknowledgement Sequence Number (ACK_SEQ)	3
	Message Sequence Number (MSG_SEQ)	3
	Acknowledgement Required Indicator (ACK_REQ)	1
	Valid Acknowledgement Indicator (VALID_ACK)	1
	Acknowledgement Address Type (ACK_TYPE)	3
	Personal Station Identifier Field Type (MSID _TYPE)	3
	Personal Station Identifier Field Length (MSID_LEN)	4
	Personal Station Identifier (MSID)	8xMSID_LEN
Message Body Field	Authentication Mode (AUTH_MODE)	2
	Authentication Data (AUTHR)	0 or 18
	Random Challenge Value (RANDC)	0 or 8
	Call History Parameter (COUNT)	0 or 6
	Registration Type (REG_TYPE)	4
	Slot Cycle Index (SLOT_CYCLE_INDEX)	3
	Protocol Revision of the Personal Station (MOB_P_REV)	8
	Extended Station Class Mark (EXT_SCM)	1
	Reserved Bits (RESERVED)	1
	Slotted Mode (SLOTTED_MODE)	1
	Reserved Bits (RESERVED)	5
	Pers. Station Terminated Calls Accepted Indicator (MOB_TERM)	1
	Reserved Bits (RESERVED)	6

Figure 4.23 The format of the message body field of the *Page Response Message*

Subfield Name	Subfield Length (bits)
Message Type (MSG_TYPE)	8
Acknowledgement Sequence Number (ACK_SEQ)	3
Message Sequence Number (MSG_SEQ)	3
Acknowledgement Required Indicator (ACK_REQ)	1
Valid Acknowledgement Indicator (VALID_ACK)	1
Acknowledgement Address Type (ACK_TYPE)	3
Personal Station Identifier Field Type (MSID _TYPE)	3
Personal Station Identifier Field Length (MSID_LEN)	4
Personal Station Identifier (MSID)	8xMSID_LEN
Reserved Bits (RESERVED)	2
Authentication Challenge Response (AUTHU)	18
Reserved Bits (RESERVED)	4

Figure4.24 The format of the message body field of the *Authentication Challenge Response Message*

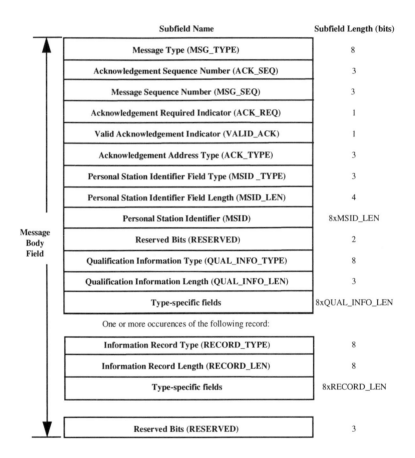

	Subfield Name	Subfield Length (bits)
	Message Type (MSG_TYPE)	8
	Acknowledgement Sequence Number (ACK_SEQ)	3
	Message Sequence Number (MSG_SEQ)	3
	Acknowledgement Required Indicator (ACK_REQ)	1
	Valid Acknowledgement Indicator (VALID_ACK)	1
	Acknowledgement Address Type (ACK_TYPE)	3
	Personal Station Identifier Field Type (MSID _TYPE)	3
	Personal Station Identifier Field Length (MSID_LEN)	4
	Personal Station Identifier (MSID)	8xMSID_LEN
Message Body Field	Reserved Bits (RESERVED)	2
	Qualification Information Type (QUAL_INFO_TYPE)	8
	Qualification Information Length (QUAL_INFO_LEN)	3
	Type-specific fields	8xQUAL_INFO_LEN

One or more occurences of the following record:

Information Record Type (RECORD_TYPE)	8
Information Record Length (RECORD_LEN)	8
Type-specific fields	8xRECORD_LEN

Reserved Bits (RESERVED)	3

Figure 4.25   The format of the message body field of the *Status Response Message*

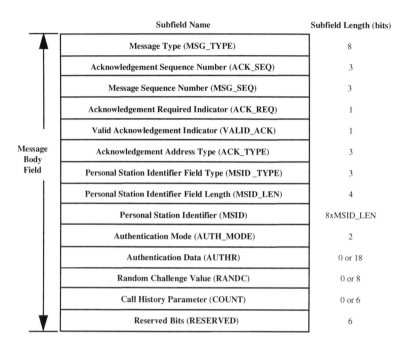

Subfield Name	Subfield Length (bits)
Message Type (MSG_TYPE)	8
Acknowledgement Sequence Number (ACK_SEQ)	3
Message Sequence Number (MSG_SEQ)	3
Acknowledgement Required Indicator (ACK_REQ)	1
Valid Acknowledgement Indicator (VALID_ACK)	1
Acknowledgement Address Type (ACK_TYPE)	3
Personal Station Identifier Field Type (MSID _TYPE)	3
Personal Station Identifier Field Length (MSID_LEN)	4
Personal Station Identifier (MSID)	8xMSID_LEN
Authentication Mode (AUTH_MODE)	2
Authentication Data (AUTHR)	0 or 18
Random Challenge Value (RANDC)	0 or 8
Call History Parameter (COUNT)	0 or 6
Reserved Bits (RESERVED)	6

Figure 4.26 The format of the message body field of the *TMSI Assignment Completion Message*

Signaling on
the Reverse Traffic Channel

The format of the reverse traffic channel message

Figure 4.27 shows the format of a reverse traffic channel message. Each message contains the following three fields:

- Message length field: This field is eight bits long, and indicates the length in octets of the reverse channel message, i.e., the sum of the lengths of the three fields. The maximum length of an access channel message is 2040 bits or 255 octets.

- Message body field: This field contains the main body of the message. The length of this field is variable ranging from 16 bits to 2016 bits.

■   CRC field: This field contains a 16-bit CRC computed for the reverse traffic
    channel message. The CRC is computed for the message length and message
    body fields.

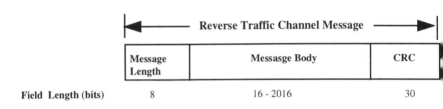

Figure 4.27 The format of an reverse traffic channel message

The message body field contains various subfields which depend on the message.
There are 17 different signaling messages transmitted on the reverse traffic channel:
the 12 messages transmitted on the reverse traffic channel only and the five which are
transmitted on both the access and reverse traffic channels.

4.7.2 Signaling on the Forward Link

Signaling on
the Synchronization Channel

A synchronization channel message to be transmitted is put into a synchronization
channel message capsule. If the message does not fill the capsule length completely,
it is appended with a padding. This synchronization channel message capsule is then
subjected to the frame formation process discussed earlier. The number of frames
formed is a multiple of three to fill a superframe of three frames. To form the
frames, the message capsule is broken into an integer number which is a multiple of
three, say 3N, of segments and each segment is appended to the SOM-field of length
one bit.

As discussed in the frame formation process, the SOM-field is set to 1 if the corresponding frame is the beginning of the message; otherwise, it is set to 0. The result is a series of 3N synchronization channel frames, each 32 bits in length and 26.66.. ms in duration. Since the SOM-field length is one bit and the synchronization channel frame length is 32 bits, the synchronization channel message capsule length must be (32-1)x(3N) bits, or 93xN bits, where N is the number of superframes. Figure 6.18 shows the relationships between a synchronization channel message, a synchronization channel message capsule, synchronization channel frames and an synchronization channel superframe.

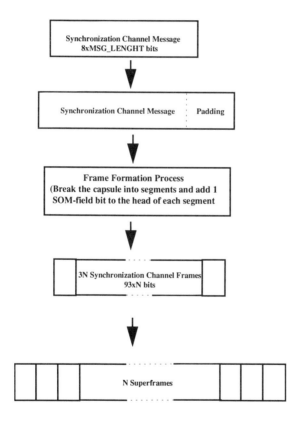

Figure 4.28 The construction of synchronization channel superframes

The format of the synchronization channel message is similar to that of the access and reverse traffic channel messages shown in Figures 6.16 and 6.17. The synchronization channel carries one type of message, *the Synchronization Channel Message*. It contains three fields: the message length field, which is eight bits long, the message body field, which is 2 - 1146 bits long, and the CRC field, which is 30 bits long. The message body field is divided into 13 subfields including the system identification (SID) field, the pilot PN sequence offset (PILOT_PN) field, the long code state (LC_STATE) field, and the system time (SYS_TIME) field.

Signaling on the Paging Channel

The paging channel is divided into 80 ms slots. Each slot repeats itself in a cycle. The maximum period of the slot cycle is 2048 slots or 163.84 seconds A paging channel message to be transmitted is put into a paging channel message capsule. If the capsule is a unsynchronized capsule, no padding is required; if is a synchronized capsule, and if the message does not fill the capsule length completely, it is appended with a padding. This paging channel message capsule is then subjected to the frame formation process discussed earlier. In order to fit into a paging channel slot, which is 80 ms long, eight half frames, each 10 ms in duration, are formed from a paging channel message. First, the paging channel message capsule is eight segments. Each segment is called a paging channel half frame body and is appended to the SCI-field of length one bit to form a paging channel half frame. These eight paging channel half frames are put into one paging channel slot. Figure 4.14 shows the process of forming paging channel half frames.

The format of the paging channel messages is similar to that of the access and reverse traffic channel messages shown in Figures 4.18 and 4.27. It contains three fields: the message length field, which is eight bits long, the message body field, whose length is variable, and the CRC field, which is 30 bits long. The message body field contains various subfields depending on the message type transmitted on the paging channel. The following is a complete list of the 20 signaling messages transmitted on the paging channel:

- System Parameters Message
- Access Parameters Message
- Reserved for Obsolete Neighbor List Message
- CDMA Channel List Message
- Reserved for Obsolete Slotted Page Message
- Reserved for Obsolete Page Message

- Order Message
- Channel Assignment Message
- Data Burst Message
- Authentication Challenge Message
- SSD Update Message
- Feature Notification Message
- Extended System Parameters Message
- Extended Neighbor List Message
- Status Request Message
- Service Redirection Message
- General Page Message
- Global Service Redirection Message
- TMSI Assignment Message
- Null Message

Signaling on the Forward Traffic Channel

The format of the forward traffic channel signaling messages is similar to that of the access and reverse traffic channel messages shown in Figures 4.18 and 4.27. It contains three fields: the message length field, which is eight bits long, the message body field, whose length ranges from 16 bits to 2016 bits, and the CRC field, which is 16 bits long. The message body field contains various subfields depending on the message type transmitted on the forward traffic channel. The following is a complete list of the 22 signaling messages transmitted on the forward traffic channel:

- Order Message
- Authentication Challenge Message
- Alert with Information Message
- Data Burst Message
- Reserved for Obsolete Hand-off Direction Message
- Analog Hand-off Direction Message
- In-Traffic System Parameters Message
- Neighbor List Update Message
- Send Burst DTMF Message
- Power Control Parameters Message
- Retrieve Parameters Message
- Set Parameters Message
- SSD Update Message
- Flash with Information Message

- Mobile Station Registered Message
- Status Request Message
- Extended Hand-off Direction Message
- Service Request Message
- Service Response Message
- Service Connect Message
- Service Option Control Message
- TMSI Assignment Message

5

HIGHLIGHTS OF OTHER WIRELESS ACCESS STANDARDS

This chapter highlights the remaining five wireless access standards for the 2 GHz PCS. They are:

- The high tier TDMA system based on the European TDMA cellular standard referred to as the Global System for Mobile (GSM)

- The Personal Access Communications System (PACS)

- The Wideband CDMA system

- The composite CDMA/TDMA system

- The DECT-based system

Sections 5.1 through 5.5 discuss the systems for the licensed PCS operation; and Section 5.6 discusses the systems for the unlicensed PCS operation.

5.1 THE GSM-BASED TDMA SYSTEM

This system is a high tier TDMA system based on the European digital cellular standard called the Global System for Mobile (GSM)[7]. The GSM was originally developed for 900 MHz and was later up-shifted in frequency to 1.8 GHz under the name of DCS1800. Forty two nations have adopted the GSM-based standard.

One fundamental difference between this system and the two North American systems reviewed in Chapters 3 and 4 is that this system specifies the total end-to-end network whereas the other two systems specify only the air interface. In this system, the inter-system operation is based on the use of its unique protocol referred to as the GSM Mobile Application Part (MAP). In the two North American systems, the inter-system operation is based on the protocol specified elsewhere referred to as the IS-41 MAP.

5.1.1 Services

The services provided by this system have been developed based on the ISDN concept with the appropriate modifications and restrictions imposed by the fact that the access to the system is over the air using the wireless technology.

Voice Services

Basic Telephony

This system provides the Mobile-Terminated call and the Mobile-Originated call handling capabilities.

Emergency Call Services

The subscriber may press the emergency button or alternatively dial "911" to access the emergency call service. Channel assignments with a number of different priority classes are possible. The allocation request for the emergency call channel assignment may pre-empt an existing connection. The emergency call is indicated by setting the calling party's category to "PRIORITY" in the outgoing Initial Address Message. In the case where there is no idle outgoing trunk available to handle the emergency call, the emergency call switching center performs several retries of the call setup.

Supplementary Services

A large number of supplementary services are available in this system. The following lists some of them:

- Call forwarding

- Call transfer

- Call waiting

- Three-way calling

- Conference calling

- Call barring

- Closed user group

- Calling number identification presentation/restriction

- Reverse charging

- Advice of charge

Voice Messaging Support

This system provides the message waiting indication capability using the SMS and the voice message retrieval capability.

Data Services

Bearer Services

The following bearer data services are supported by this system:

- Asynchronous data circuits

  - Circuit-switched data at 300, 1200, 2400, 4800 and 9600 b/s

  - Packet Assembler Dissembler (PAD) access at 300, 1200, 2400, 4800, and 9600 b/s

- Synchronous data circuits

  - Circuit-switched data at 1.2, 2.4, 4.8 and 9.6 kb/s

  - Packet data at 2.4, 4.8 and 9.6 kb/s

- Alternate speech/data

- Speech followed by data

Data Connection Types

This system supports both the transparent connection and the non-transparent connection.

The transparent connection. In this type of connection, the Layer 2 protocol is used end-to-end between the MS and the fixed network terminal equipment. No additional error protection is provided other than the forward error correction employed. There is no buffering of information in the Interworking Function (IWF) device. This type of connection provides a constant delay via the air interface.

The non-transparent connection. In this type of connection, the Layer 2 protocol is terminated in the MS and in the IWF. Additional error protection is provided over the air by means of a radio link protocol based on the Automatic Repeat Request (ARQ) technique. This type of connection results in a variable delay via the air interface.

Interworking to the fixed network for the data services. The following interworking capabilities are provided for the data services:

- *3.1 kHz audio connection.* Modems are used for the data transmission by converting. The analog modem signals are converted to the 64 kb/s trunk signals for the MSC-to-fixed network connections.

- *Unrestricted digital, circuit switched connections.* The user data rate is adapted to the 64 kb/s trunk rate by means of the rate adaptation techniques according to that used in the IWF and that used in the ISDN terminal.

■ *Unrestricted digital connections for the packet calls.* The 64 kb/s unrestricted digital trunks are used. The rate adaptation is performed by the X.31 flag stuffing technique.

Short Message Services

Two types of Short Message Services (SMS's) are provided by this system:

■ Point-to-Point SMS

■ Broadcast SMS

Point-to-Point SMS. The point-to-point SMS is provided between the MS and the Short Message Service Center (SM-SC), which performs the store-and-forward functions on the messages to deliver them to their ultimate destinations. Short messages up to a maximum of 140 octets can be exchanged. The SMS data transfer occurs in parallel to the established speech or data calls using the signaling channels, the Standalone Dedicated Channel (SDCCH) and the Slow Associated Channel (SACCH). Both the Mobile-Originated and the Mobile-Terminated point-to-point SMS services are provided. Interworking between different networks such as the PSTN, the ISDN and the Local Area Networks (LAN's) as well as other services such as the X.400 based services are possible.

The SM-SC supports non-mobile entities as well as the mobile entities. Therefore, the mobile user can send short messages to and receive short messages from the non-mobile entities, for example, for the paging services and the e-mail services.

Cell Broadcast SMS. This service allows broadcasting of short messages up to a maximum of 82 octets within selected cells. The Teletext service on TV is an example. Up a maximum of 15 messages may be concatenated to form a macro message.

The Dual-Tone
Multi-Frequency Signaling

The Dual-Tone Multi-Frequency (DTMF) signaling is an in-band signaling technique. The support of this signaling method is useful when the system must inter-work with other networks where no out-of-band signaling such as the SS7 signaling is used and the DTMF signaling is used instead. The DTMF signaling is supported in this system by using special messages. When the mobile subscriber depress a DTMF button on the keypad, say 'x,' this information is passed on to the MSC via the BSS in a DTMF message. The MSC applies the appropriate DTMF tone on the outgoing facility such as the trunk to the PSTN. After applying the tone on the outgoing facility, the MSC sends a DTMF acknowledgment message to the subscriber's MS.

5.1.2 The Network Architecture

Figure 5.1 shows the end-to-end network architecture of the GSM-based 1.9 GHz system.

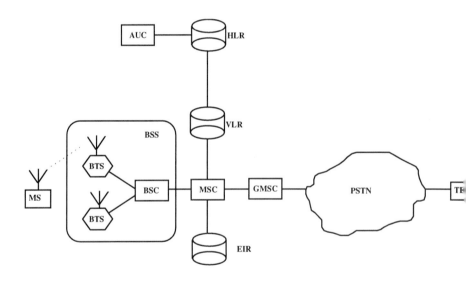

Figure 5.1 The end-to-end network architecture of the GSM-based 1.9 GHz system

The network elements shown in Figure 5.1 are described below:

■ *Home Location Register (HLR)*. The subscriber is assigned with a unique HLR. The subscriber registers the location in the HLR. The HLR also stores other relevant information regarding the subscriber's profile.

■ *Visitor Location Register (HLR)*. The VLR stores the location and other relevant information about the subscriber who is currently roaming in its area. The Mobile Switching Center (MSC) retrieves this information to handle the calls for the subscriber. The VLR data is updated by the HLR when a subscriber enters the area. The HLR stores the location of the VLR of the subscriber's current location.

■ *Mobile Switching Center (MSC)*. The MSC is the heart of the system and processes the calls to and from the mobile station. The MSC, in conjunction with the VLR, provides the following functions: the standard switching functions between all of its trunk interfaces; the SS7 signaling functions with the Mobile Application part (MAP) and the ISDN User Part (ISUP); country specific signaling towards the PSTN/ISDN; the mobile radio functions; Operations and Maintenance functions; and charging data collection functions.

■ *Gateway MSC (GMSC)*. The GMSC is the MSC handling the interface between the cellular network and the Public Switched telephone Network (PSTN).

■ *Authentication Center (AUC)*. The AUC provides the other network entities, namely, the HLR and the VLR, with the parameters required in the authentication process such as the triplets of Challenge, Response and Communications Key. The AUC is used to prevent handling of the secret keys and algorithms from being spread throughout the network.

■ *Base Station System (BSS)*. The BSS refers to a collection of equipment which make up the Base Station as seen by the Mobile Switching Center (MSC) and consists of

  – Base Transceiver Station Controller (BTS): The BTS is similar to the system element referred to as the Base Station (BS) in the two North American systems discussed in Chapters 3 and 4. The BTS consists of the radio equipment and other equipment to support a single cell. The BTS performs the following specific functions: rate adaptation; channel coding and decoding; interleaving and de-interleaving; frame formation; encryption and decryption; modulation/demodulation; and radio link measurement.

  – Base Station Controller (BSC): The BSC controls a number of BTS's. Specifically, the BSC performs the following functions: radio resource management; power control management; handoff management, and management of interfaces.

■ *Equipment Identity Register (EIR).* The EIR is a data base which handles the registration of the approved mobile terminals and can be used, for example, to blacklist stolen terminals.

5.1.3 Privacy and Authentication

Privacy

This system provides an encryption capability using a secret key algorithm in which the session keys are changed during the authentication procedure. The encryption capability is provided for the following types of confidentiality:

■ *User information confidentiality for speech and data*

■ *Signaling information confidentiality*

■ *Subscriber identity confidentiality*

■ *Subscriber location confidentiality.* This is provided by using the International Mobile Subscriber Identity (IMSI) and the Temporary Mobile Subscriber Identity (TMSI). The IMSI is not published and is not sent over the air interface. The TMSI is used for the identification of the MS and is used within the local VLR.

Authentication

In this system, the authentication procedure is performed for the subscriber's MS at the time of an initial location registration, a call set-up, or an exchange of short messages. The authentication procedure is also performed at the time of a connectionless supplementary service, a location update when the VLR is changed or when a handoff occurs.

The authentication mechanism is based on the use of the Personal Identity Number (PIN), the SIM card, and a complex challenge-response algorithm. The user enters the PIN, and the SIM checks the PIN locally. After the mobile station passes this local check, it must then pass the authentication procedure by interacting with the network as follows as depicted in Figure 5.2. The network generates a non-predictable random number called RAND and sends it to the MS as a challenge. In response to this challenge, the MS executes the authentication function and computes the signature of the received RAND called SRES (Signed RESult) using a certain algorithm specified in this system and a secret key specific to the user denoted in this system by *Ki*. The MS transmits the resulting SRES to the network. In parallel to this, the network also performs its authentication function using the same RAND and the same authentication key *Ki* unique to this particular MS to compute SRES. The network tests the SRES received from the MS and the SRES computed by itself for validity. If the two SRES's agree, the MS's authentication is successful and the MS is accepted; otherwise, the MS is rejected.

The execution of the authentication function also generates the communication key both by the MS and by the network. If the authentication is successful, the network and the MS enter the ciphering mode using this communication key and encrypt the user data and signaling information.

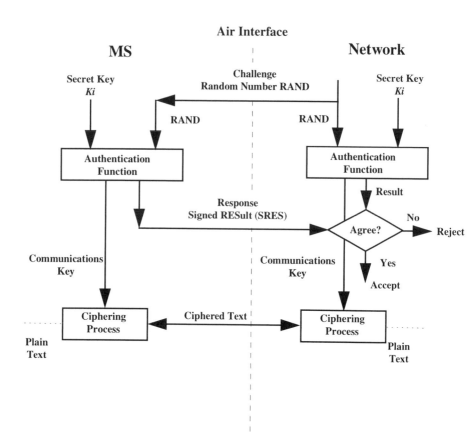

Figure 5.2 The authentication and encryption procedures

The Subscriber Identity Module

A unique feature about this system is that the mobile station is divided into two parts, the Mobile Equipment (ME) and the Subscriber Identity Module (SIM). The SIM stores all the subscriber-related information; and the ME does not have any subscriber identity. When the SIM is inserted into the ME, the two together becomes the mobile station for the user. This feature allows portability of the mobile station.

The user only needs to carry the SIM card and can use it, for example, with a rental car as long as the car is equipped with an ME (i.e., the other half of the mobile station).

The SIM card protects the subscriber and the service provider. It stores the following information about the subscriber:

- Personal Identity Number (PIN)

- Temporary Mobile Subscriber Identity (TMSI)

- Ciphering key, Ki

- Authentication algorithm which is service-provider-proprietary

- Communication key computation algorithm which is service-provider-proprietary

5.1.4 Handoff

A handoff is possible between the physical channels or cells controlled by the same BSS, between the BSS's controlled by the same MSC, and between the BSS's belonging to different MSC's of the same network. The choice of a new channel is done by the network based on the measurements performed by the MS of the signal strength and quality of the current channel and of the signal strength of the surrounding base stations. The latter measurements are done by the MS on time slots other than the current channel. The MS sends the measurements to the MSC via the Slow Associated Channel (SACCH). Since the measurements are made by the MS, this type of handoff is called mobile-assisted handoff. The initial MSC serves as the anchor MSC and is in charge of the complete call including any subsequent handoff to a third MSC.

5.1.5 The Logical Channel Structure

The logical channels in this system are divided into two categories: traffic channels (TCH's) and control channels (CCH's). The traffic channels and the control channels are further divided into various logical channels as shown in Figure 5.3.

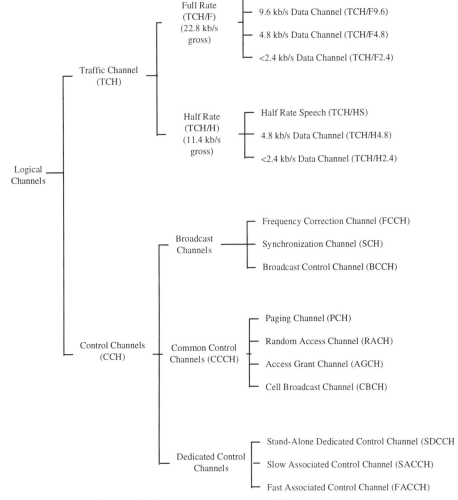

Figure 5.3   The logical channels in the GSM-based system

The Traffic Channels (TCH) carry the user speech or data. It operates at the gross bit transmission rate of 22.8 kb/s for the full rate channel and 11.4 kb/s for the half rate channel. The Broadcast Channels are transmitted continuously by the Base Stations.   These channels carry the synchronization information and the system information required by the MS's. The Common Control Channels (CCCH) are used in both the down link direction, i.e., the Paging Channel (PCH) and the Access Grant

Channel (AGCH), and the uplink direction, i.e., the Random Access Channel (RACH). Finally, the Dedicated Control Channels (DCCH) are for a specific MS.

5.1.6 Creation of Physical Channels

In this system, the physical channels are created by the combination of the FDD, the FDMA, and the TDMA. In addition, the frequency hopping method is used for the diversity.

The Relationship between the FDD, FDMA and TDMA

First, the total bandwidth is divided into the up-link and the down-link bands. Each of the two bands is then further divided into 200-kHz-wide bands to create frequency channels. Each 200-kHz frequency channel in each direction is then used to create time slots by the TDMA technique. The frequency hopping method is employed to provide the multipath frequency diversity and the interferer diversity.

Figure 5.4 shows the FDD, the FDMA and the TDMA for this system.

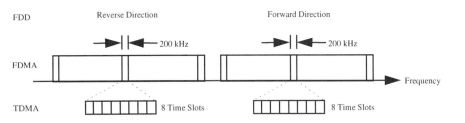

Figure 5.4 The FDD, the FDMA and the TDMA

The Frame Structure

Figure 5.4 shows the frame structure of this system. The duration of a TDMA frame is 4.615 milliseconds. One TDMA frame contains 1250 bit positions (BP's). Therefore, the maximum bit transmission rate in each transmission direction is 270.8559 kb/s and the inter-BP time is 3.692 µs. Each frame is divided into eight time slots. The PCS spectrum of 120 MHz yields 60 MHz for each direction. Since the channel spacing is 200 kHz, 60 MHz yields 300 frequency channels in each direction. Since each frequency channel yields eight time slots, a total of 2400 time slots are possible in each direction, i.e., 2400 full duplex channels, from the 120 MHz PCS spectrum.

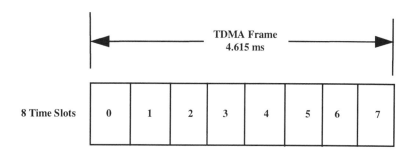

Figure 5.5   The frame structure

The Time Slot Structure

A basic radio channel is a time slot. The duration of one time slot is 576.875 µs. Since each frame contains 1250 bit positions (BP's), one time slot contains 156.25 BP's. All time slots have this same duration and this same number of BP's. However, the structure of the time slot varies depending on the type of the data burst. Five different types of time slot data bursts are defined as follows:

■ *Normal Burst (NB)*. Carries information on the Traffic Channels and the Control Channels.

■ *Frequency Correction Burst (FB)*. This burst is used to synchronize the frequency of the MS's. It generates the Frequency Correction Channel (FCCH).

■ *Synchronization Burst (SB)*. It is used for the synchronization of the MS's. It carries the TDMA frame number and the base station identity code (BSIC). It generates the Synchronization Channel (SCH).

■ Dummy Burst (DB). This is used when no data is required to be transmitted.

■ Access Burst (AC). This is used by the MS to access the system on the Random Access Channel (RACH) and is also used after a handoff.

Figure 5.5 shows the time slot structures for the five burst types. For the first four burst types, out of 156.25 BP's, only 148 BP's constitute the burst and the remaining 8.25 BP's are left as a time gap until the next time slot as a guard time. For the Normal Burst, 114 bits per time slot are used for the user data. Since these bits repeat every frame duration, which is 4.615 ms, the bit transmission rate of a basic traffic channel is 24.7 kb/s in each transmission direction. The guard time is 8.25 BP duration, which is 30.459 μs. The Access Burst has 36 bits for the user data, and, therefore, the corresponding bit transmission rate is 7.8 kb/s. The guard time for the Access Burst is 252 μs.

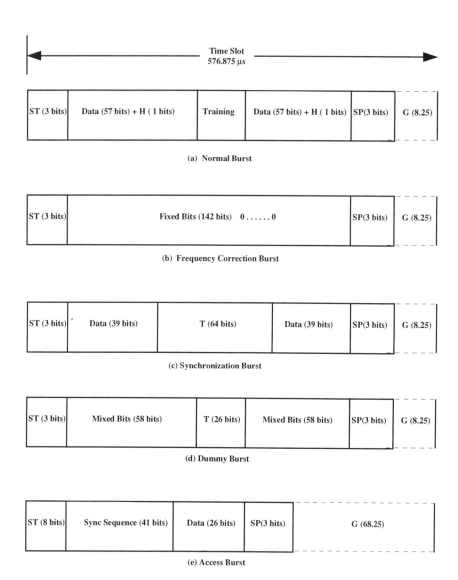

ST = Start Bits SP = Stop Bits H = Stealing Bits T Training Bits G = Guard Time

Figure 5.6 The time slot structures

The Multiframe, Superframe and Hyperframe Structures

A multiframe consists of 26 TDMA frames lasting 120 ms. A superframe consists of 51 multiframes lasting 6.12 seconds. A hyperframe consists of 2048 superframes lasting 3.48 hours. Figure 5.7 illustrates the relationship between the TDMA frame, the multiframe, the superframe and the hyperframe for the traffic and associated control channels.

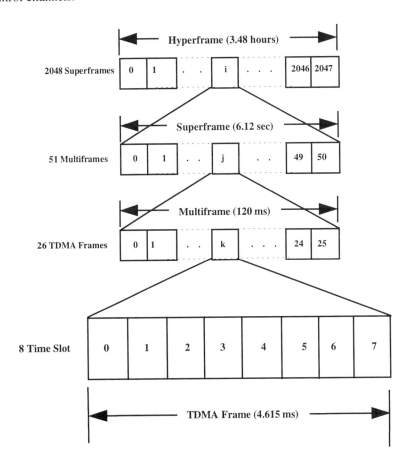

Figure 5.7 The frame, multiframe, superframe and hyperframe for traffic/associated control channels

The multiframe structure shown in Figure 5.7 is used to multiplex the traffic channel and the associated control channels. Out of 26 TDMA frames contained in on multiframe, 24 frames are used for the traffic channels and one frame is used for th Slow Associated Control Channels (SACCH's), and one frame is left idle. Eac traffic channel frame contains eight full rate individual channels, i.e., time slo Since each time slot is repeated in each traffic frame in a multiframe, each specif time slot is repeated 24 times in one multiframe duration, which is 120 ms (4.615 n x 26 = 119.99 ms to be exact). Recall that each normal burst time slot contains 11 information bits. Hence, the bit transmission rate of each full rate individual chann is computed to be 22.8 kb/s as follows:

Total number of information bits transmitted in one multiframe
duration by one individual channel

= (114 bits)(24 frames)

= 2736 bits

Bit transmission rate for one individual channel

= (Number of bits transmitted in one multiframe) / (Multiframe duration)

= (2736 bits) / (120 ms)

= 22.8 kb/s

There are eight SACCH's, which repeat themselves every multiframe. Eacl SACCH, therefore, transmits 114 bits in every 120 ms yielding the bit transmissior rate of 950 b/s.

Half rate channels repeat every other frame and the bit rate is 11.4 kb/s. The SACCH's for the half rate channels have the same bit rate, 950 b/s.

Figure 5.8 illustrates the relationship between the TDMA frame, the multiframe, the superframe and the hyperframe for the non-associated control channels.

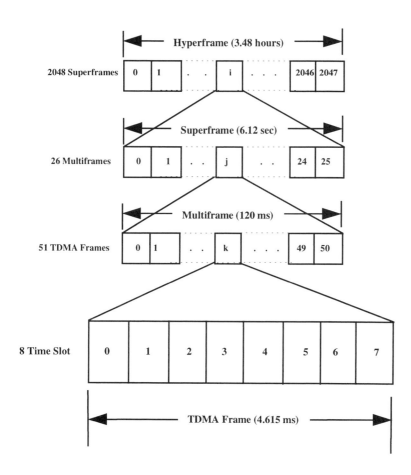

Figure 5.8 The relationship between the TDMA frame, the multiframe, the superframe and the hyperframe for the non-associated control channels

5.1.7 Speech Coding, Channel Coding, Interleaving and Modulation

Speech Coding

This system uses the Regular Pulse Excitation with Long Term Prediction (RP LTP) speech coding for the full rate. The speech signal is sampled at the rate 8000 samples per second. Each sample is coded in 13 bits. This bit stream processed by the speech coding algorithm to produce 260 bits of 20 ms frame. T result is a digital binary source bit stream at the rate of 13 kb/s for the full ra channel.

Channel Coding

This system uses the block code for the error detection and the convolutional codi with the code rate r = 1/2 for the forward error correction.

Interleaving

This system uses speech frame interleaving.

Modulation

This system uses the Gaussian Minimum Shift Keying (GMSK) with a Bandwidt Time product (BT-product) of 0.3.

5.2 THE PERSONAL ACCESS COMMUNICATIONS SYSTEM

The Personal Access Communications System (PACS) is a low tier TDMA system, which is new and is not based on any existing standard[3].

5.2.1 Summary of System Parameters

The following list summarizes some of the system parameters of the PACS, which will be discussed in subsequent subsections in more detail:

Physical channel creation	FDD/FDMA/TDMA
RF channel spacing	300 kHz
Frame duration	2.5 ms
No. of time slots/frame	8
Time slot duration	312.50 µs
No. of bits per time slot	120 bits
No. of bits per fram	960 bits
Inter-bit position time	2.6 µs
Bit rate per RF channel	384 kb/s
Gross bit rate per time slot	48 kb/s
Speech coding	32 kb/s ADPCM
Channel coding	Error detection (12-bit CRC)
Modulation	$\pi/4$-QPSK
MS transmit power	100 mW peak
BS transmit power	800 mW peak
Power control	Uplink (yes); downlink (no)

5.2.2 The PACS System Architecture

Figure 5.9 shows the PACS system architecture.

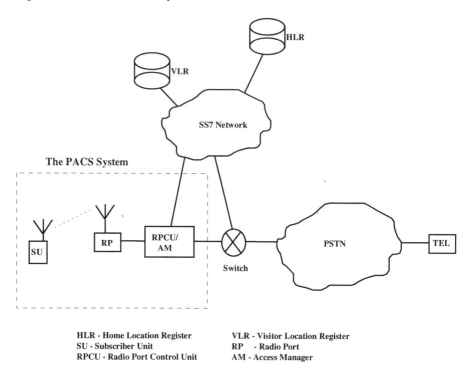

HLR - Home Location Register   VLR - Visitor Location Register
SU - Subscriber Unit   RP   - Radio Port
RPCU - Radio Port Control Unit   AM - Access Manager

Figure 5.9   The PACS system architecture

5.2.3 Creation of Physical Channels

In this system, the physical channels are created by the combination of the FDD, the FDMA, and the TDMA.

The Relationship between the FDD, FDMA and TDMA

First, the total bandwidth is divided into the up-link and the down-link bands. Each of the two bands is then further divided into 300-kHz-wide bands to create frequency channels. Each 300-kHz frequency channel in each direction is then used to create time slots by the TDMA technique. Figure 5.10 illustrates the FDD, the FDMA and the TDMA methods used in this system to create physical channels.

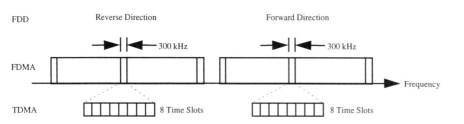

Figure 5.10   The FDD, the FDMA and the TDMA

The Frame Structure

Figure 5.11 shows the frame structure of this system. The duration of a TDMA frame is 2.5 ms. One TDMA frame contains 960 bit positions (BP's). Therefore, the maximum bit transmission rate in each transmission direction is 384 kb/s and the inter-BP time is 2.6 μs. Each frame is divided into eight time slots. Hence, the gross bit rate of each time slot is 48 kb/s.

The PCS spectrum of 120 MHz yields 60 MHz for each direction. Since the channel spacing is 300 kHz, 60 MHz yields 200 frequency channels in each direction. Since each frequency channel yields eight time slots, a total of 1600 time slots are possible in each direction, i.e., 1600 full duplex channels, from the 120 MHz PCS spectrum.

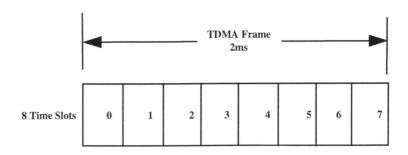

Figure 5.11 The PACS frame structure

The Time Slot Structure

A basic radio channel is a time slot. The duration of one time slot is 312.50 µs. Since each frame contains 960 bit positions (BP's) and there are eight time slots per frame, one time slot contains 120 BP's. All time slots have this same duration and this same number of BP's. However, the structure of the time is slightly different between the uplink and the downlink. Figure 5.12 shows the time slot structures of the down link and the uplink.

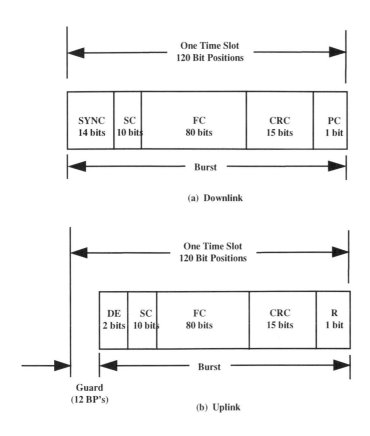

SYNC - Synchronization   SC - Slow Channel   FC - Fast Channel   PC - Power Control
DE - Differential Encoder   R - Reserved

Figure 5.12   The PACS time slot structures

The time slot structures shown Figure 5.12 define the following physical channels:

■ *The Synchronization Channel (SYC).* The SYNC channel is present on the downlink only and is not present on the uplink. Fourteen bits are assigned for the SYNC channel. Since each time slot repeats itself every 2.5 ms, 14 bits of the SYNC channel repeats themselves every 2.5 ms, yielding the SYNC channel bit rate of 5.6 kb/s.

■ *The Slow Channel (SC).* The SC is present on both downlink and the uplink. Ten bits are allocated for the SC to operate at 4 kb/s: The SC contains information about:

- Whether the current channel is the SBC, Idle TC or Busy TC
- Whether the current channel is the Message Channel (MC) or the Synchronous Directive Channel (SDC)
- Word Error Indicator (WEI) bit
- Bandwidth of the SBC (8, 6, or 32 kb/s)
- Available bandwidth and the access type permitted for the idle TC
- Message type carried in the Fast Channel (FC)

■ *The Fast Channel (FC).* The FC is present on both downlink and the uplink. Eighty bits are allocated for the FC to operate at 32 kb/s. The FC carries the Traffic Channel or the System Broadcast Channel (SBC) as will be discussed later.

■ *Cyclic Redundancy Check (CRC).* Fifteen bits are assigned for this function on both directions of transmission.

■ *Power Control.* One bit is assigned for this function on the downlink. The power control is not used on the uplink, and the corresponding bit position is reserved for future use.

5.2.4 The Logical Channel structure

There are two types of time slots. one type of time slot contains, among others, the Traffic Channel (TC), and the other type contains, among others, the System Broadcast Channel (SBC). Out of the eight time slots per frame, seven times slots (Time Slot 0, 1, 2, 3, 4, 6 and 7) contain the TC, and one time slot (Time Slot 5) contains the SBC. The time slot that contains the SBC, Time Slot 5, is active at all times. The other seven time slots containing the TC may be idle or busy.

The relationship between the physical channels and the logical channels is discussed below. The time slot containing the SBC, i.e., Time Slot 5, contains the SBC as the only logical channel. The SBC is carried in the FC of that time slot. The SBC has three sub channels: the System Information Channel (SIC), the Alerting Channel (AC) and the Priority Request Channel (PRC). The time slot which contain the TC may contain two other logical channels: the Message Channel (MC) and the

Synchronous Directive Channel (SDC). The TC is carried in the FC of that time slot. The logical channels are summarized below:

- *The System Broadcast Channel (SBC)*

    - System Information Channel (SIC). This channel carries the system information in the downlink.

    - Alerting Channel (AC). The channel carries alerting messages in the downlink.

    - Priority Request Channel (PRC). This channel is present on both the downlink and the uplink. Since the word *broadcast* in the channel name *SBC* implies the BS-to-the MS direction, i.e., the downlink, the SBC may sound as though it is unidirectional. However, it is bi-directional and, in fact, contains the PRC, which is bi-directional. The PRC carries priority link access requests on the uplink and responds to them on the downlink.

- *The Traffic Channel (TC)*

- *Message Channel (MC)*

- *Synchronous Directive Channel (SDC)*

- *User Information Channel (UIC)*

On the downlink, all of the above logical channels are available. On the uplink, however, only the TC, the MC, the SDC, the UIC and the PRC are available.

5.2.5 Messages

Layer 2 Messages

The following lists Layer 2 messages used in the PACS:

Message Type	Message Name
Channel Access	INITLAL_ACCESS
	ACCESS_CONFIRM
	ACCESS_DENY
	ACCESS_RELEASE
L2 System Broadcast	PRIORITY_ACCESS_REQ
	PRIORITY_ACCESS_ACK
	BLOCKED_ACCESS
	BROADCAST_ALERT
	SYSTEM_INFO
L2 Error & Flow Control for L3	ACK_MODE_TRANS
	INFO_ACK
	RECEIVER_NOT_READY
	RECEIVER_READY
Automatic Link Transfer (ALT)	LINK_SUSPEND
	LINK_RESUME
	TST_REQ
	ALT_REQ
	ALT_ACK
	ALT_DENY
	ALT_EXEC
	ALT_COMP
	PERFORM_ALT
	PERFORM _TST
Maintenance and Testing	LOOPBACK
	DEFEAT_ANTENNA_DIV
	RESUME_ANTENNA_DIV
	INITIATE_TEST_MODE
	CLEAR_TEST_MODE

The ACK_MODE_TRANS Message Format

Among the messages listed above, ACK_MODE_TRANS is unique in that it is used to carry Layer 3 messages and provides an error control capability with the Go-Back-N ARQ protocol whereas the remaining messages are used for Layer 2 signaling.

One ACK_MODE_TRANS message is eight-octets long and completely fits into the Fast Channel (FC) of a time slot, which has 80 bit positions. Layer 3 messages can be encapsulated in one ACK_MODE_TRANS or, if it is too long to fit into one, can be broken into up to four segments.

The ACK_MODE_TRANS message containing the beginning segment of a Layer 3 message begins with a two-octet L2 header and ends with the L3 message segment; the ACK_MODE_TRANS message containing an intermediate segment of a Layer 3 message begins with a one-octet continuation header and ends with the L3 message segment; and the ACK_MODE_TRANS message containing the last segment of a Layer 3 message begins with a one-octet continuation header and ends with a two-octet check sum at the end of the message.

Figure 5.13 shows the content of the ACK_MODE_TRANS message for the four possible cases of the Layer 3 message segmentation: the case of no segmentation and the cases of two segments, three segments, and four segments. Also shown in Figure 5.13 are the total number of octets of a Layer 3 message that can be carried in each case, which are four, 11, 18 and 25 octets with one, two three and four ACK_MODE_TRANS messages, respectively. In other words, in order to fit into one ACK_MODE_TRANS message, the length of the Layer 3 message must be no greater than four octets because of the two-octet header and the two-octet checksum. The maximum length of a Layer 3 message that can be transmitted in a sequence of ACK_MODE_TRANS message (i.e., four of them) is 25 octets or 200 bits.

No. of Octets of L3 Message

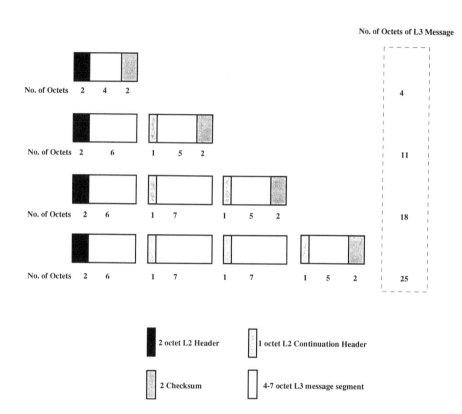

Figure 5.13   ACK_MODE_TRANS structure for Layer 3 message segmentation

Figure 5.14 shows the formats of the two-octet L2 header and the one-octet L2 continuation header.

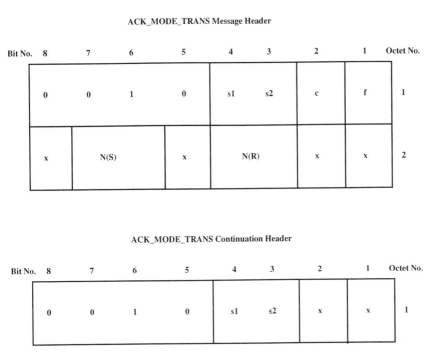

Figure 5.14 ACK_MODE_TRANS header formats

The Error Control Using the ACK_MODE_TRANS Message

The PACS uses the Go-Back-N ARQ scheme for the error control. In this scheme, a certain window of acknowledgments is defined. The receiver checks the frames received within this window, and, if any frame is missing or corrupted, it lets the send know about that frame. The send must go back to that frame, say N frames back, and retransmit all the N past frames from that missing (or corrupted) frame to the current frame.

N(S) and N(R) are the send and receive message numbers which are kept track of in the ARQ protocol. The acknowledgment for an ACK_MODE_TRANS message can be done either directly by sending an L2 message INFO_ACK or by piggybacking on a returning ACK_MODE_TRANS message.

The Flow Control

The PACS provides a flow control mechanism. When the receiving end needs to tell the sending end to stop sending frames, it sends an L2 message RECEIVER_NOT_READY; and when it wants to have the sender resume the transmission, it sends an L2 message RECEIVER _READY.

Layer 3 Messages

The following lists Layer 3 messages used in the PACS:

Message Name

RCID_REQ
RCID_ACK
RELEASE
REL_COM
CALL_REQ
ALERT_ACK
INCOMING_CALL
CONNECT
PRIORITY_CALL_REQ
PRIORITY_CALL_ACK
TERM_REG_REQ
TERM_REG_ACK
INFO

5.2.6 Authentication

The authentication procedure used in this system is essentially the same as that used in the IS-136-based system discussed in Chapter 3, which is a private key method. This system also supports migration to the use of a public key method.

5.2.7 The Automatic Link Transfer Procedures

In this system, handoff is referred to as the Automatic Link Transfer (ALT). In this system, the MS selects a new channel in addition to performing link measurements. The network performs only switching function when the new channel is selected. Four distinct cases of ALT are considered in this system

- *Time Slot Transfer (TST)*. ALT between two time slots

- *Intra RPCU ALT*. This is the ALT between the Radio Ports (RP's) on the same Radio Port Control Unit (RPCU).

- *Inter-RPCU ALT*. This is the ALT between the RP's connected to different RPCU's on the same switch.

- *Inter-switch ALT*. This is the ALT between RP's connected to different RPCU's on different switches.

While a call is in progress, the user's Subscriber Unit (SU) continually monitors the channel quality of the surrounding Radio Ports (RP's) and compares it with its current Traffic Channel (TC). When the SU finds a better TC on another RP, the SU enters into the ALT procedure as depicted in Figure 5.15.

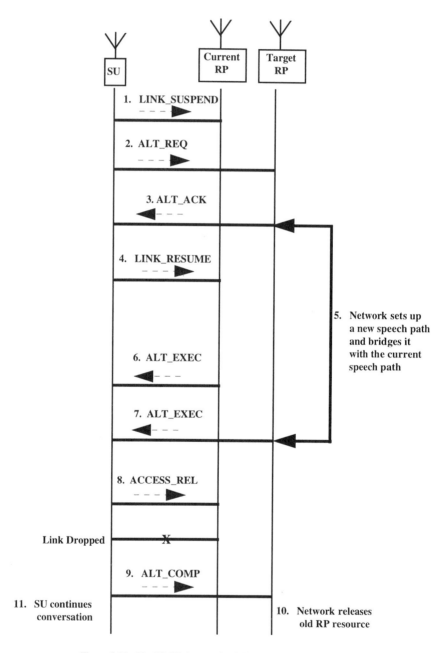

Figure 5.15 The PACS Automatic Link Transfer (ALT) procedure

1. The SU suspends the current TC by sending a Layer 2 (L2) message LINK_SUSPEND. The SU then searches for a TC with a channel rate compatible with the current TC.

2. The SU informs the target RP of its request for an ALT by sending an L2 message ALT_REQ.

3. The target RP acknowledges the request by sending an L2 message ALT_ACK to the SU.

4. The SU sends an L2 message LINK_RESUME to the target RP and returns to the original RP to resume the current call.

5. While the SU is continuing with the current call, the target RP and the network sets up a new speech path to an idle TC on the target RP and the current speech path and the new speech path are bridged.

6. Once the new speech path is set up, the network orders the SU to perform the ALT by sending an L2 message ALT_EXEC via the current RP.

7. The network sends the ALT_EXEC message to the SU via the target RP also.

8. Upon receiving this order, the SU sends an L2 message ACCESS_RELEASE to the current RP and moves over to the new RP dropping the current link.

9. After moving over to the new RP, the SU sends an L2 message ALT_COMP to the new RP.

10. The network releases the radio resource associated with the old RP.

11. The SU continues the current conversation.

An ALT may be initiated by the network also. To start a network initiated ALT, the network sends an L2 message PERFORM_ALT to the SU.

5.2.8 Data Services

Data Interworking

The PACS provides the following data interworking capabilities:

- *Interworking with the PSTN.* The interworking with the PSTN is provided in two ways. First, the digital data from the user's data device is converted to an analog signal through a modem. The analog signal is then inputted into the MS for source coding. The MS transmits the resulting signal as it does with the speech signal. The other method is to transmit the digital data signal directly over the air, and let the Interworking device in the network convert the digital signal into the analog signal for the PSTN transmission. Figure 5.16 illustrates these two methods of interworking with the PSTN.

- *Interworking with the ISDN.* This interworking is done using the rate adaptation for the 64-kb/s multiples of digital lines.

- *Interworking with the Packet Switched Public Data Network (PSPDN).* This interworking is done by using the X.25 protocols and the Packet Assembler/Dissembler (PAD) protocols X.3, X.28 and X.29.

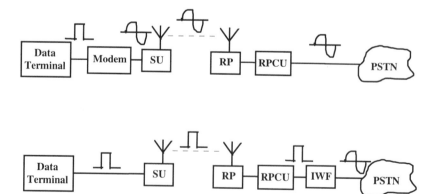

Figure 5.16   The PACS data interworking

The Link Access Protocol for Radio

The PACS provides the Link Access Protocol for Radio (LAPR) for the interworking with the PSTN and the ISDN. For the PSTN interworking, the LAPR is applied between the PACS's Subscriber Unit (SU) and the Interworking Function (IWF) device in the network. The IWF provides the interface between the PACS and the PSTN. For the ISDN interworking, the LAPR is applied between the PACS SU and the PACS radio Port Control Unit (RPCU). The RPCU then provides the interface between the PACS and the ISDN network.

The LAPR protocol provides the framing, synchronization and the ARQ capabilities similar to those provided by the ISDN Q.921 LAPD protocol. Figure 5.17 shows the LAPR frame structure.

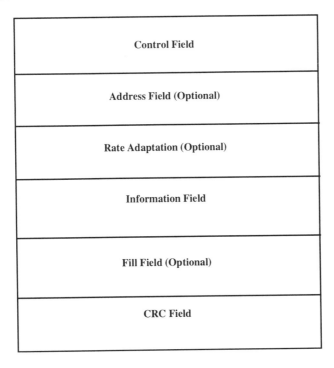

Figure 5.17 The PACS LAPR frame structure

The LAPR protocol operates in the following three modes:

- *The acknowledged mode.* In this mode, the LAPR provides the Go-Back-N ARQ operation.

- *The unacknowledged mode.* In this mode, the LAPR provides the transparent data transmission.

- *The bit-synchronization mode.* The operation uses a network independent clock and bit-level sequence numbers.

5.3   WIDEBAND CDMA SYSTEM

This system is not based on any existing standard and is a new system[8]. The following are some of the main characteristics of the Wideband CDMA system:

- *Bandwidth.* The frequency channel bandwidth is 5 MHz.

- *Logical channel structure.* Although the channel structures are different, the same four types of the logical channels used in the IS-95-based system discussed in Chapter 4 are used in this system as well: the Access Channel, the Reverse Traffic Channel, the Paging Channel and the Forward Traffic Channel.

- *Multiple access.* This system uses the CDMA method.

- *Authentication.* The authentication procedure used in this system is essentially the same as that used in the IS-136-based system discussed in Chapter 3.

- *Handoff.* This system provides the following two types of handoff: mobile station assisted handoff and base station assisted handoff. The mobile station assisted handoff has the following two procedures: idle handoff and in-traffic handoff. The mobile station initiates the handoff procedure when the pilot strength from other base stations exceeds the specific threshold or the transmission power of the mobile station exceeds the specific maximum transmission power. The base station assisted handoff is initiated when any of the following conditions exists: transmission power of the designated mobile station exceeds the threshold, or cell loading exceeds the threshold, or distance based option.

5.4 COMPOSITE TDMA/CDMA SYSTEM

This system is new and is not based on any existing standard. The following are some of the main characteristics of this system:

- *Multiple access.* This system uses combination of TDMA, FDMA and CDMA. Within a PCS cell, TDMA is used to separate users. To provide greater area of coverage or to provide greater capacity for densely populated regions, multiple cells or sectorized cells are deployed using FDMA. Finally, to permit multi-cell deployments in a given region, the direct sequence spread spectrum CDMA is used for each RF link to reduce co-channel interference between cells using the same RF carrier frequency.

- *Authentication.* This system supports an authentication procedure similar to that of the DCS-1800-based system and alternatively to one similar to that of the IS-136-based system, both of which are discussed earlier.

- *Handoff.* This system uses a mobile station controlled handoff method. While the mobile station is receiving bearer traffic from its originating base station, it measures the received signal quality of its link. This value, together with the current frame error, determines the link quality. If the quality drops below a first threshold, the mobile station searches during the TDMA time slots it is not using, for other RF frequencies/PN code sets through all time slots of the originating base station. As the mobile station finds each new frequency/code set, it measures the received signal quality. The mobile station reads a field carried in all base station frames which describes the current time slot utilization of the base station. The mobile station uses these two pieces of information to form a figure of merit. When the link quality drops a second threshold level, the mobile station requests a handoff from the base station.

5.5 DECT-BASED SYSTEM

TAG-6 decided to pursue the wireless access standards for the unlicensed operation only and is working with TR-41. Therefore, this system will be discussed as part of Section 5.6.

5.6 WIRELESS ACCESS STANDARDS FOR THE UNLICENSED OPERATIONS

As discussed in Section 1.2, the FCC also allocated frequency bands for the unlicensed PCS operation. In these frequency bands, the PCS operator does not need to obtain a license. The typical environment for the unlicensed PCS operation is inside a building. Wireless PBX/Centrex and wireless Local Area Network are some of the potential applications of the unlicensed PCS operation.

The following systems are for the unlicensed PCS operation:

1. *The PACS-U(A)[4]*. This system is included in the PACS standard as Annex A.

2. *The PACS-U(B)[5]*. This system is included in the PACS standard as Annex B.

3. *The Personal Communications Interface (PCI)*. This system is based on the existing CT2+ system.

4. *The North American/Wireless Customer Premises Equipment (NA/WCPE)*. This system is based on the existing Digital European Cordless Telephone (DECT) system

5. *The Orthogonal CDMA system*. This is a new system.

The systems for the unlicensed operation of PCS must satisfy the following rules specified by the FCC, which are commonly referred to as the "etiquette:"

1. Coordination rule
2. Listen-before-talk rule
3. Thirty-second rule
4. Channelization rule
5. Packing rule
6. Power level rule
7. Frame duration rule

The Coordination Rule

The unlicensed PCS system must be coordinated with the fixed microwave radio system for the frequency spectrum. The following is a quote from the FCC Etiquette Rule, Part III, Section C, Paragraph 91: "We also require that manufacturers demonstrate that any movable components (i.e., handsets or terminals) of the device or system will be prevented from transmitting if those parts leave the coordinated area around the base station."

The Listen-Before-Talk Rule

The following is a quote from the FCC Etiquette Rule, Part III, Section C, Paragraph 15.323: "(1) Immediately prior to transmission, devices must monitor the combined time and spectrum windows in which they intend to transmit for a period of at least 10 milliseconds to determine if the access criteria are met. (2) The monitoring threshold must not be more than 30 dB above the thermal noise power for a bandwidth equivalent to the emission bandwidth used by the device. (3) If no signal above the threshold level is detected, transmission may commence and continue with the same emission bandwidth in the monitored time and spectrum windows without further monitoring."

The Rule further states that: "An initiating device that is prevented from monitoring during its intended transmit window due to monitoring system blocking from the transmissions of a co-located (within one meter) transmitter of the same system, may monitor the portions of the time and spectrum windows in which they intend to receive over a period of at least 10 milliseconds. The monitored time and spectrum window must total at least 50 percent of the 10 millisecond frame interval and the monitored spectrum must be within the 1.25 MHz frequency channel(s) already occupied by that device or collocated co-operating devices. If the access criteria is met for the intended receive time and spectrum window under the above conditions, then transmission in the intended transmit window by the initiating device may commence."

The Thirty-Second Rule

The following is a quote from the FCC Etiquette Rule, Part III, Section C, Paragraph 15.323: "Once access to specific combined time and spectrum window is obtained an acknowledgment from a system participant must be received by the initiating transmitter within one second or transmission must cease. Periodic acknowledgments must be received at least every 30 seconds or transmission must cease. Channels used exclusively for control and signaling information may transmit continuously for 30 seconds without receiving an acknowledgment, at which time the access criteria must be repeated."

The Channelization Rule

The FCC Etiquette Rule, Part III, Section 15.323 (a) specifies a maximum channelization of 1.25 MHz in the isochronous unlicensed spectrum from 1920 to 1930 MHz and the minimum channelization of 50 kHz.

The Packing Rule

The following is a quote from the FCC Etiquette Rule, Part III, Section 15.323 (b): "Intentional radiators with an intended emission bandwidth less than 625 kHz shall start searching for an available time and spectrum window within 3 MHz of the sub-band edge at 1920 MHz and search upward from that point. Devices with an intended emission bandwidth greater than 625 kHz shall start searching for an available time and spectrum window within 3 MHz of the sub-band edge at 1930 MHz and search downward from that point."

The Power Level Rule

The following is a quote from the FCC Etiquette Rule, Part III, Section 15.319 (c): "Peak transmit power shall not exceed 100 microwatts multiplied by the square root of the emission bandwidth in hertz. Peak transmit power must be measured over any interval of continuous transmission using instrumentation calibrated in terms of an rms-equivalent voltage. The measurement results shall be properly adjusted for any instrument limitations, such as detector response times, limited resolution bandwidth capability when compared to the emission in question over the full bandwidth of the channel."

The Frame Duration Rule

The following is a quote from the FCC Etiquette Rule, Part III, Section 15.319 (e): "The frame period (a set of consecutive time slots in which the position of each time slot can be identified by reference to a synchronized source) of an intentional radiator operating in these sub-bands shall be 20 milliseconds/X where X is a positive whole number.

REFERENCES

[1] N. S. Jayant and P. Noll, *Digital Coding of Waveforms*, Prentice-Hall.

[2] J-STD-008, "Personal Station-Base Station Compatibility Requirements for 1.8 to 2.0 GHz for Licensed and Unlicensed Applications."

[3] J-STD-014, "Personal Access Communications System (PACS) Air Interface Standard."

[4] Supplement to J-STD-014, "Personal Access Communications System Unlicensed (version A) Air Interface Standard."

[5] Supplement to J-STD-014, "Personal Access Communications System Unlicensed (version B) Air Interface Standard."

[6] J-STD-011, "PCS 1900 MHz IS-136-based Air Interface Compatibility Standard."

[7] J-STD-007, "Air Interface Specification for 1.8 to 2.0 GHz Frequency Hopping Time Division Multiple Access (TDMA) for Personal Communications Service."

[8] J-STD-015, "Wideband CDMA Air Interface Compatibility Standard for 1.8 to 1.99 GHz Personal Communications Service."

ABOUT THE AUTHOR

Dr. Kun Il Park has 25 years experience in the telecommunications field including 14 years of technical management experience. From 1973 to 1984, Dr. Park worked for Bell Laboratories on various network planning assignments initially as Member of Technical Staff and then as Supervisor of the Data Communications Network Planning Group. He joined Bell Communications Research, Inc. (Bellcore) with the Bell System divestiture in 1984, and, since then, has held various positions as follows: District Manager of the ISDN Planning Group responsible for network planning for ISDN and Broadband ISDN, ATM congestion control, and network performance standards; Director of the Network Operations Analysis Group responsible for software requirements for network operations systems; and currently Professional Services Consultant responsible for network planning and systems engineering in the area of personal and wireless communications and mobility management. He has participated in the standards activities of Committee T1, the National Bureau of Standards, the Corporation for Open Systems, and most recently the Joint Technical Committee for PCS wireless access standards.

Dr. Park has published numerous papers for conferences and journals including the Bell System Technical Journal, the IEEE Transactions on Communications, and the IEEE Transactions on Vehicular Technology. He has given many technical seminars on wireless and personal communications including a one-day seminar entitled "Wireless and Personal Communications" as Invited Lecturer at Monmouth University, West Long Branch, New Jersey. Since 1985, he has been Adjunct Professor of Electrical Engineering at the Stevens Institute of Technology, Hoboken, New Jersey, and, over the years, has taught various subjects including the graduate courses entitled "Wireless Communications," "Digital Communications Engineering," "Probability and Stochastic Processes," and "Analytical Methods in Electrical Engineering." He has one patent on a multi-tier PCS system currently awaiting approval.

Dr. Park received his B.S. degree in electrical engineering from Seoul National University and his M.S. and Ph.D. degrees in electrical engineering from the University of Pennsylvania. He is Senior Member of IEEE .

INDEX

—2—

2 GHz, 177

—3—

3.1 kHz, 180

—9—

900 MHz, 177

—A—

A-band, 4, 5, 95, 96
Access channel, 131, 132, 134, 147, 150, 152, 155, 156, 157, 158, 159, 160, 161, 162, 163, 171
Access Response Channel, 55, 59
Acknowledged mode, 59, 60, 95, 100, 214
Adaptive DPCM, 20, 22
ADPCM, 20, 22, 23, 27, 106, 197
Advanced Mobile Phone Service, 27, 55
Air, 1, 5, 6, 33, 38, 46, 50, 53, 57, 62, 67, 178, 180, 184, 212
Air interface, 1, 5, 6, 46, 50, 53, 57, 62, 67, 178, 180, 184, 219

ALT, 204, 209, 210, 211
AM, 33, 35
Amplitude modulation, 33
AMPS, 27, 55
Analog, 9, 10, 11, 12, 13, 15, 27, 33, 55, 105, 106, 124, 175, 180, 212
ARCH, 55, 59, 95, 98, 100
Architecture, 182, 198
ARQ, 59, 83, 84, 90, 91, 95, 96, 97, 98, 99, 100, 101, 102, 122, 180, 205, 207, 213, 214
Asynchronous, 4, 179
Asynchronous data, 179
Auction, 1, 5
Authentication, 56, 115, 116, 117, 124, 160, 169, 175, 183, 184, 185, 186, 187, 208, 214, 215
Authentication Center, 183
Automatic Repeat Request, 99, 180

—B—

Base Station, 38, 42, 119, 158, 183, 184, 188, 219
Base Station Controller, 184
Base Station System, 183
Base Transceiver Station Controller, 183

Basic Trading Area, 5
B-band, 5, 56, 95, 97, 117
BCCH, 55, 59, 60, 71, 72, 73, 75, 78, 80,
    81, 91, 92, 93, 94, 98
Bit position, 40, 47, 49, 50, 52, 67, 68,
    70, 76, 94, 100, 102, 103, 108, 190,
    197, 199, 200, 202, 205
Bit rate, 35, 38, 39, 40, 45, 46, 47, 50, 51,
    53, 73, 194, 197, 199, 201
Bit transmission rate, 20, 21, 23, 24, 25,
    27, 35, 37, 39, 40, 46, 49, 50, 52, 53,
    76, 77, 78, 80, 81, 104, 109, 136, 188,
    190, 191, 194, 199
Block coding, 28, 30
Blocking, 217
Broadcast Control Channel, 55, 59
BS, 38, 42, 43, 49, 57, 68, 99, 100, 102,
    105, 115, 116, 117, 119, 120, 183,
    197, 203
BSC, 184
BSS, 182, 183, 187
BTA, 5
BTS, 183, 184
Burst, 31, 84, 98, 104, 109, 157, 158,
    160, 166, 175, 190, 191, 194

—C—

Call processing, 56, 154, 155, 157, 158
C-band, 4, 5
CDMA, 2, 6, 7, 8, 45, 53, 123, 124, 125,
    126, 127, 128, 154, 155, 156, 158,
    174, 177, 214, 215, 216, 219
Cell, 3, 6, 183, 214, 215
Cell size, 3
Cellular, 1, 2, 3, 5, 6, 7, 8, 27, 55, 123,
    124, 177, 183
CELP, 25, 106

Channel coding, 12, 13, 15, 28, 39, 62,
    68, 71, 73, 105, 107, 146, 147, 183,
    196, 197
Channel Spacing, 190, 197, 199
Channelization rule, 216, 218
Chip rate, 53, 130, 152, 153
Circuit-switched, 180
Code channel, 128, 129, 130, 154, 155
Code Division Multiple Access, 2, 6, 45,
    53, 124
Code Excited Linear Predictive coding,
    25
Code rate, 30, 73, 109, 120, 122, 147,
    196
Committee T1, 5
Companding, 19, 106
Composite TDMA/CDMA, 215
Convolutional coding, 28, 196
Coordination rule, 216
CRC, 28, 29, 31, 71, 84, 86, 90, 108, 135,
    163, 172, 174, 175, 197, 202
CT2+, 216
Cyclic Redundancy Check, 28, 86, 135,
    202

—D—

Data, 4, 13, 15, 28, 29, 30, 56, 68, 84, 85,
    88, 93, 103, 104, 106, 107, 109, 110,
    115, 116, 117, 118, 120, 130, 131,
    132, 146, 147, 151, 155, 157, 158,
    160, 166, 175, 179, 180, 181, 183,
    184, 185, 188, 190, 191, 212, 214, 221
Data class, 107
Data rate, 106, 155, 180
D-band, 4, 5
DCCH, 55, 56, 59, 60, 68, 103, 189
DCS1800, 177

DECT, 7, 8, 177, 215, 216
Delta Modulation, 20, 22, 23
Differential PCM, 20, 21
Digital, 1, 7, 8, 9, 10, 11, 12, 13, 14, 15,
    16, 19, 20, 27, 39, 55, 56, 59, 60, 69,
    70, 103, 104, 105, 106, 108, 109, 116,
    146, 177, 180, 181, 196, 212, 216, 219
Digital communications, 9, 10, 11, 12,
    13, 15, 16
Digital Control Channel, 55, 56, 59, 60,
    69, 70
Digital European Cordless Telephone, 7,
    216
Digital Traffic Channel, 56, 103, 116
Directory Number, 112
DM, 20, 22, 23, 24
DN, 112, 113
Down-link, 197, 200, 201, 202, 203
DPCM, 20, 21, 22, 23
DQPSK, 37, 38, 109
DTC, 56, 103, 104, 120, 122
DTMF, 160, 175, 181
Duplexing, 14, 42, 43, 45, 65, 124

—E—

E-band, 4, 5
EIR, 184
Electronic Serial Number, 110, 113, 158
Encryption, 13, 99, 184, 186
Equipment Identity Register, 184
Error, 12, 13, 16, 19, 20, 21, 22, 28, 31,
    37, 99, 107, 180, 196, 197, 202, 204,
    205, 207, 215
Error control, 28, 99, 205, 207
Error correction, 28, 180, 196
Error detection, 28, 31, 196
ESN, 110, 113, 116, 119, 158

Etiquette, 216
Extended Broadcast Control Channel, 55

—F—

Fast Broadcast Control Channel, 55
F-band, 4
FCC, 1, 2, 3, 4, 5, 38, 65, 115, 124, 216,
    217, 218
FDCCH, 55, 59, 68, 70, 71, 73, 75, 76,
    77, 78, 80, 81, 82, 83, 91
FDD, 43, 45, 46, 47, 48, 50, 51, 53, 65,
    66, 124, 125, 189, 197, 198, 199
FDMA, 45, 46, 47, 65, 66, 124, 125, 189,
    197, 198, 199, 215
Federal Communications Commission, 1,
    115
Flow control, 204, 208
FM, 33, 34, 35, 38, 140, 141, 147
Forward Digital Control Channel, 59, 70
Forward link, 47, 65, 66, 67, 68, 124,
    125, 126, 128, 130, 131, 147, 149,
    151, 153, 172
Frame Check Sequence, 29
Frame duration, 47, 49, 50, 52, 53, 75,
    103, 106, 132, 133, 134, 136, 144,
    145, 148, 149, 191, 197, 216
Frame duration rule, 216, 218
Frame format, 67, 71, 72, 83, 86, 87, 134,
    144, 145, 146, 147, 161, 172, 173,
    174, 184, 190, 199, 200, 213
Frame formation, 146, 147, 161, 172,
    173, 174, 184
Frame length, 40, 83, 92, 103, 132, 133,
    134, 136, 145, 148, 161, 173
Frequency channel, 65, 66, 67, 125, 126,
    127, 128, 129, 130, 146, 189, 190,
    199, 214, 217

Frequency Division Duplexing, 43, 65, 124
Frequency Division Multiple Access, 45, 46, 65, 66, 124, 125
Frequency modulation, 33, 34
Full-rate, 67, 188, 194, 196

—G—

Gateway MSC, 183
Gaussian Minimum Shift Keying, 38, 196
Global System for Mobile, 6, 177
GMSC, 183
GMSK, 38, 196
Gray code, 37
Gray coding, 37
GSM, 6, 8, 177, 178, 182, 188

—H—

Hamming distance, 30
Handoff, 56, 124, 184, 185, 187, 191, 209, 214, 215
High-tier, 177
HLR, 119, 120, 183
Home Location Register, 183
Hyperframe, 83, 93, 193, 194, 195

—I—

IMSI, 84, 85, 110, 156, 184
Indication primitive, 57
Interleaving, 28, 31, 32, 62, 68, 71, 73, 107, 109, 146, 149, 184, 196
International Mobile Subscriber Identity, 184
Interworking, 180, 181, 212, 213
Interworking Function, 180, 213
ISDN, 178, 180, 181, 183, 212, 213, 221

ISDN User Part, 183
Isochronous, 4, 218
ISUP, 183
IWF, 180, 213

—J—

Joint Technical Committee, 6, 221
JTC, 6, 7

—L—

LAPD, 213
LAPR, 213, 214
Layer 2, 57, 58, 61, 62, 63, 68, 71, 72, 73, 74, 75, 83, 86, 87, 91, 93, 94, 180, 204, 205, 211
Layer 3, 57, 61, 62, 64, 71, 72, 83, 84, 85, 91, 93, 94, 95, 99, 100, 102, 205, 206, 208
License, 4, 216
Licensed operation, 3, 4
Linear Predictive Coding, 15, 25, 106
Link Access Protocol for Radio, 213
Logical channel, 55, 56, 58, 59, 60, 65, 68, 70, 71, 75, 78, 79, 80, 81, 83, 91, 99, 131, 132, 133, 134, 146, 147, 148, 149, 151, 153, 159, 187, 188, 202, 203, 214
Long code, 130, 153, 155, 174
Loss, 12, 16
Low-tier, 197
LPC, 15, 25, 27

—M—

MAHO, 56
Major Trading Area, 5
MAP, 91, 100, 102, 178, 183

Mapping, 37, 61, 62, 63, 71, 73, 74, 75, 76, 77, 78, 81, 82, 98, 101, 108, 112, 121, 153, 154
MC, 202, 203
Mean Opinion Score, 25
Message Channel, 202, 203
Message format, 122, 205
Message segmentation, 61, 71, 205, 206
MIN, 84, 88, 110, 111, 112, 113
Mobile Application Part, 178
Mobile Assisted Handoff, 56
Mobile Identification Number, 110, 112
Mobile Station, 38, 41, 84, 85, 105, 176
Mobile Switching Center, 183
Modem, 180, 212
Modulation, 11, 13, 20, 21, 22, 23, 33, 34, 36, 37, 38, 39, 106, 109, 153, 184, 196, 197
MOS, 25, 26, 27
MS, 38, 41, 43, 49, 57, 59, 68, 84, 85, 88, 90, 98, 99, 100, 102, 104, 105, 110, 113, 114, 115, 116, 117, 119, 120, 180, 181, 182, 184, 185, 187, 188, 191, 197, 203, 209, 212
MSC, 180, 182, 183, 187
MTA, 5
Multiframe, 193, 194, 195
Multiplex, 137, 139, 140, 141, 142, 143, 194
Multiplexing, 13, 123, 135, 136, 137, 138, 140, 142

—N—

NA/WCPE, 216
Network, 1, 2, 3, 5, 6, 21, 56, 155, 178, 180, 182, 183, 185, 187, 209, 211, 212, 213, 214, 216, 221

Noise, 12, 16, 31, 53, 155, 217
North American, 2, 6, 7, 8, 19, 55, 123, 124, 178, 183, 216
North American/Wireless Customer Premises Equipment, 216
Nyquist, 16

—O—

Orthogonal CDMA, 216

—P—

Packet Assembler Dissembler, 179
Packing rule, 216, 218
PACS, 7, 8, 177, 197, 198, 200, 201, 204, 207, 208, 210, 212, 213, 216, 219
PAD, 179, 212
Paging Channel, 55, 59, 123, 132, 133, 145, 148, 174, 188, 214
PCH, 55, 59, 95, 98, 100, 188
PCI, 216
PCM, 20, 21, 22, 23, 106
PCS, 1, 2, 3, 4, 5, 7, 55, 65, 123, 124, 177, 190, 199, 215, 216, 219, 221
PCS spectrum allocations, 4
Performance, 12, 16, 21, 23, 25, 26, 27, 221
Personal Access Communications System, 7, 177, 197, 219
Personal Communications Interface, 216
Personal Handy Phone, 7
Phase modulation, 33, 34, 38
PHP, 7
Physical channel, 38, 39, 41, 42, 46, 47, 49, 50, 51, 53, 57, 58, 65, 67, 103, 124, 128, 130, 187, 189, 198, 199, 201, 202
Physical Layer, 6, 57

PM, 33, 34, 35, 36, 38
PN, 155, 158, 174, 215
Power, 3, 6, 104, 105, 113, 114, 155, 157, 159, 160, 175, 184, 197, 202, 214, 216, 217, 218
Power control, 175, 184, 197, 202
Power level rule, 216, 218
Primitive, 57
Priority Request Channel, 202, 203
Privacy, 56, 117, 184
Protocol, 57, 58, 59, 83, 84, 91, 92, 95, 96, 97, 98, 99, 100, 101, 178, 180, 205, 207, 213, 214
Pseudo noise, 155
PSTN, 181, 182, 183, 212, 213
Public Switched Telephone Network, 183
Pulse Code Modulation, 20, 21, 106

—Q—

Q.921, 213
QPSK, 36, 38, 153, 197
Quadrature Phase Shift Keying, 36, 37, 109, 153
Quantization, 10, 11, 15, 16, 17, 18, 19, 20, 21, 22, 23
Quantization error, 16, 19, 20

—R—

RACH, 55, 59, 60, 71, 72, 73, 74, 75, 83, 84, 86, 87, 88, 91, 92, 93, 99, 100, 189, 191
Radio Port, 209
Radio Port Control Unit, 209
Random Access Channel, 55, 59, 189, 191
Rate set, 131, 132, 133, 135, 136, 137, 138, 140, 141, 147, 148, 149

RDCCH, 55, 59, 68, 69, 73, 83, 91
Read matrix, 149, 151, 152
Read operation, 149, 150, 151
RELP, 25
Request primitive, 57
Residual Excited Linear Predictive coding, 25
Reverse Digital Control Channel, 59, 69
Reverse link, 65, 66, 68, 124, 126, 128, 130, 131, 147, 149, 150, 151, 152, 153, 159, 160
RP, 209, 211
RPCU, 209, 213

—S—

Sampling, 10, 15, 16, 19, 20, 21, 22, 23, 24, 33, 106
SAP, 57, 68
SBC, 202, 203
SC, 181, 202
SCF, 55, 59, 60, 70, 71, 75, 80, 81, 91, 102
SCM, 113, 114
Security, 110, 119
Service Access Point, 57, 68
Session key, 184
Shared Channel Feedback, 55, 59, 60, 70
Shared Secret Data, 56, 110, 115, 116, 117, 118
Short Message Service, 55, 56, 59, 181
Signaling, 13, 56, 57, 120, 123, 131, 137, 138, 140, 142, 157, 159, 161, 163, 171, 172, 174, 175, 181, 183, 184, 185, 205, 217
SIM, 185, 186, 187
Simplex, 42
Slow Channel, 202

SMS, 55, 59, 60, 179, 181
SMS Broadcast Control Channel, 55
SMS Channel, 55, 59
SMSCH, 55, 59, 95, 98, 100
Source coding, 15, 20, 28, 39, 212
SPACH, 55, 59, 71, 72, 73, 75, 78, 80,
    81, 83, 84, 90, 91, 92, 95, 96, 97, 98,
    99, 100, 102
Speech coding, 11, 15, 16, 20, 21, 22, 27,
    39, 105, 106, 196, 197
Spreading, 130, 146, 152, 153
SS7, 181, 183
SSD, 56, 110, 115, 116, 117, 118, 119,
    120, 175
Standard, 2, 6, 7, 8, 55, 56, 78, 106, 111,
    123, 124, 177, 183, 197, 214, 215,
    216, 219, 221
Station Class Mark, 110, 113
SU, 209, 211, 213
Subjective evaluation, 25
Subscriber Identity Module, 186
Subscriber Unit, 209, 213
Superframe, 75, 76, 77, 78, 79, 81, 82,
    93, 132, 144, 172, 173, 193, 194, 195
Supervision, 120
Symbol rate, 35, 37, 38, 39, 147, 148,
    149, 153
Symbol repetition, 146, 147
Synchronization, 131, 133, 144, 155, 172,
    174, 191, 201
Synchronization channel, 131, 132, 133,
    144, 147, 149, 151, 155, 158, 172,
    173, 174, 191, 201
Synchronous, 180, 202, 203
Synchronous data, 180
System Broadcast Channel, 202, 203

—T—

TAG, 7, 215
TDD, 43, 45, 47, 50, 51, 52
TDMA, 2, 6, 7, 8, 45, 47, 48, 49, 50, 51,
    52, 53, 55, 56, 65, 67, 69, 70, 75, 76,
    77, 81, 102, 109, 124, 177, 189, 190,
    191, 193, 194, 195, 197, 198, 199,
    215, 219
TDMA block, 75, 76, 77, 102
Technical Adhoc Group, 7
Telecommunications Industry Associate,
    5
Temporary Mobile Subscriber Identity,
    184, 187
TIA, 5
Time Division Duplexing, 42, 43, 45
Time Division Multiple Access, 2, 6, 45,
    47, 65, 67, 219
Time slot, 40, 47, 49, 50, 51, 55, 56, 62,
    63, 65, 67, 68, 69, 70, 71, 73, 74, 75,
    76, 77, 78, 80, 81, 83, 84, 85, 88, 89,
    90, 91, 103, 104, 108, 109, 120, 121,
    122, 187, 189, 190, 191, 192, 194,
    197, 199, 200, 201, 202, 205, 209,
    215, 218
TMSI, 84, 88, 156, 160, 171, 175, 176,
    184, 187

—U—

UIM, 110
Unacknowledged mode, 59, 95, 214
Unlicensed, 3, 4, 177, 215, 216, 218, 219
Up-link, 189, 197, 200, 201, 202, 203
User Identity Module, 110

—V—

Vector Sum Excited Linear Predictive
  coding, 25
Visitor Location Register, 183
VLR, 183, 184, 185
VSELP, 25, 106

—W—

WACS, 7
Waveform coding, 15
Wideband CDMA, 8, 177, 214, 219
Wireless access, 1, 2, 5, 6, 7, 38, 55, 123,
  177, 215, 221

Wireless Access Communications
  System, 7
Wireless access standards, 2, 5, 7, 177,
  215, 216, 221
Wireless PBX/Centrex, 216
Write matrix, 149, 151, 152
Write operation, 149

—X—

X.25, 212
X.400, 181